Introduction to Nanomaterials

Introduction to Nanomaterials

Jeffrey Carson

STATES
ACADEMIC PRESS
www.statesacademicpress.com

Published by States Academic Press,
109 South 5th Street,
Brooklyn, NY 11249, USA

ISBN: 978-1-63989-308-9

Cataloging-in-Publication Data

Introduction to nanomaterials / Jeffrey Carson.
p. cm.
Includes bibliographical references and index.
ISBN 978-1-63989-308-9
1. Nanostructured materials. 2. Nanocomposites (Materials). I. Carson, Jeffrey.
TA418.9.N35 I58 2022
620.115--dc23

For information on all States Academic Press publications
visit our website at www.statesacademicpress.com

STATES
ACADEMIC PRESS

Contents

Preface

Materials with individual grain size in the range of 1-100 nm are called nanomaterials. They are studied under nanotechnology, in which the basic principles are governed by materials science. Nanomaterials often have mechanical, optical and electrical properties different from those of bulk materials. This is due to the increased surface area and quantum confinement effect in nanomaterials. On the basis of their origin, nanomaterials can be classified into 3 categories: engineered, incidental and natural. The materials which have been created by humans to have some specific properties are called engineered nanomaterials. Incidental nanomaterials are those which are obtained as a by-product of some industrial process. Those which are found in biological systems are referred to as natural nanomaterials. They can also be classified as one, two or three dimensional according to the number of dimensions with components less than 100 nm in size. Some prominent examples of nanomaterials are quantum dots, computer chips, nanotubes, etc. Most of the topics introduced in this book cover new techniques and the applications of nanomaterials. While understanding the long-term perspectives of the topics, it makes an effort in highlighting their impact as a modern tool for the growth of the discipline. This book will serve as a valuable source of reference for those interested in this field.

Given below is the chapter wise description of the book:

Chapter 1- The materials which have at least one dimension in the nanoscale are known as nanomaterials. Some of the prominent types of nanomaterials are nanoparticle, nanoporous materials, nanocomposite materials and nanocrystalline materials. This is an introductory chapter which will introduce briefly all these significant aspects of nanomaterials.

Chapter 2- The structures whose size lies between microscopic and molecular structures are known as nanostructures. Some of major examples of nanostructure are quantum dot, nanowire, nanofiber, nanorod, nanoring, nanomesh, nanoflower and nanotube. This chapter discusses in detail these nanostructures.

Chapter 3- The particles of matter whose size lies between 1 and 100 nanometers are known as nanoparticles. Some of its prominent examples are magnetic nanoparticles, radioactive nanoparticle, nanosponges, nanofluid, magnetic-plasmonic bifunctional nanoparticles and nanoshell. The diverse aspects of these nanoparticles have been thoroughly discussed in this chapter.

Chapter 4- The multiphase solids which have one or more dimensions less than 100 nanometers are known as nanocomposites. A few of the major types of nanocomposites are polymer nanocomposite, carbon nanotube, silver nanoparticle and nanowire. All these diverse types of nanocomposites have been carefully analyzed in this chapter.

Chapter 5- Nanometrology focuses on the science of measurement at the nanoscale level. It plays an important role in the production of nanomaterials and devices accurately within the field of nanomanufacturing. This chapter has been carefully written to provide an easy understanding of the varied facets of nanometrology as well as nanometrology measurement techniques.

At the end, I would like to thank all those who dedicated their time and efforts for the successful completion of this book. I also wish to convey my gratitude towards my friends and family who supported me at every step.

Jeffrey Carson

1

Understanding Nanomaterials

The materials which have at least one dimension in the nanoscale are known as nanomaterials. Some of the prominent types of nanomaterials are nanoparticle, nanoporous materials, nanocomposite materials and nanocrystalline materials. This is an introductory chapter which will introduce briefly all these significant aspects of nanomaterials.

Nanomaterials

Nanomaterials are cornerstones of nanoscience and nanotechnology. Nanostructure science and technology is a broad and interdisciplinary area of research and development activity that has been growing explosively worldwide in the past few years. It has the potential for revolutionizing the ways in which materials and products are created and the range and nature of functionalities that can be accessed. It is already having a significant commercial impact, which will assuredly increase in the future.

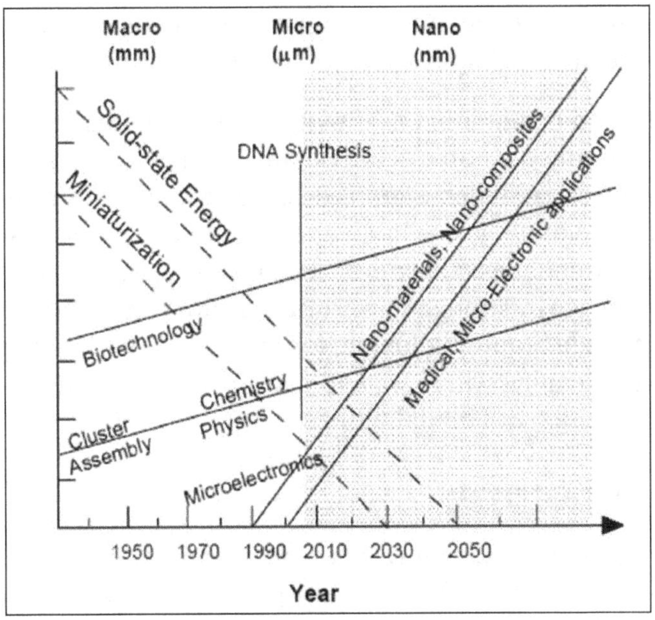

Evolution of science & technology and the future.

Nanoscale materials are defined as a set of substances where at least one dimension is less than approximately 100 nanometers. A nanometer is one millionth of a millimeter approximately 100,000 times smaller than the diameter of a human hair. Nanomaterials are of interest because at this scale unique optical, magnetic, electrical, and other properties emerge. These emergent properties have the potential for great impacts in electronics, medicine, and other fields.

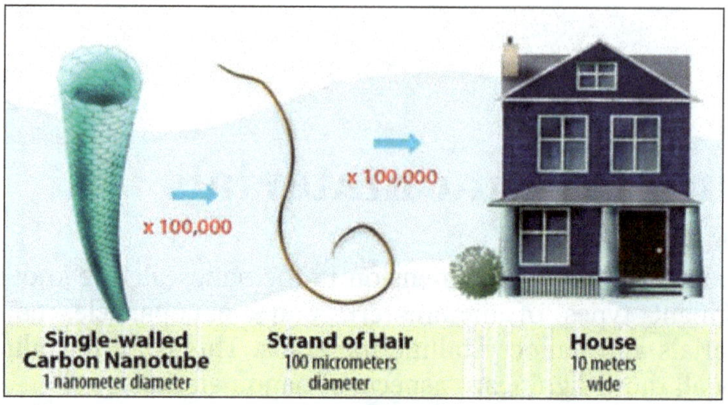

Carbon nanotube.

Some nanomaterials occur naturally, but of particular interest are engineered nanomaterials (EN), which are designed for, and already being used in many commercial products and processes. They can be found in such things as sunscreens, cosmetics, sporting goods, stain-resistant clothing, tires, electronics, as well as many other everyday items, and are used in medicine for purposes of diagnosis, imaging and drug delivery. Engineered nanomaterials are resources designed at the molecular (nanometre) level to take advantage of their small size and novel properties which are generally not seen in their conventional, bulk counterparts. The two main reasons why materials at the nano scale can have different properties are increased relative surface area and new quantum effects. Nanomaterials have a much greater surface area to volume ratio than their conventional forms, which can lead to greater chemical reactivity and affect their strength. Also at the nano scale, quantum effects can become much more important in determining the materials properties and characteristics, leading to novel optical, electrical and magnetic behaviours.

Nanomaterials are already in commercial use, with some having been available for several years or decades. The range of commercial products available today is very broad, including stain-resistant and wrinkle-free textiles, cosmetics, sunscreens, electronics, paints and varnishes. Nanocoatings and nanocomposites are finding uses in diverse consumer products, such as windows, sports equipment, bicycles and automobiles. There are novel UV-blocking coatings on glass bottles which protect beverages from damage by sunlight, and longer-lasting tennis balls using butyl-rubber/nano-clay composites. Nanoscale titanium dioxide, for instance, is finding applications in cosmetics, sun-block creams and self-cleaning windows, and nanoscale silica is being used as filler in a range of products, including cosmetics and dental fillings.

Classification of Nanomaterials

Nanomaterials have extremely small size which having at least one dimension 100 nm or less. Nanomaterials can be nanoscale in one dimension (e.g. surface films), two dimensions (e.g. strands or fibres), or three dimensions (e.g. particles). They can exist in single, fused, aggregated or agglomerated forms with spherical, tubular, and irregular shapes. Common types of nanomaterials include nanotubes, dendrimers, quantum dots and fullerenes. Nanomaterials have applications in the field of nano technology, and displays different physical chemical characteristics from normal chemicals (i.e., silver nano, carbon nanotube, fullerene, photocatalyst, carbon nano, silica). Nanostructured materials are classified as Zero dimensional, one dimensional, two dimensional, three dimensional nanostructures.

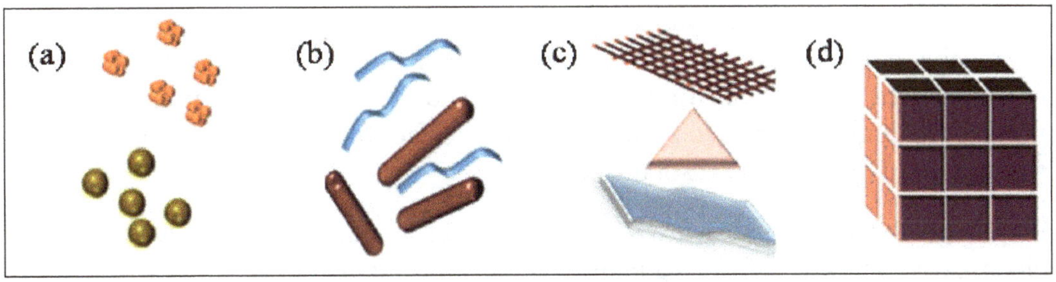

Classification of Nanomaterials (a) 0D spheres and clusters; (b) 1D nanofibers, nanowires, and nanorods; (c) 2D nanofilms, nanoplates, and networks; (d) 3D nanomaterials.

Nanomaterials are materials which are characterized by an ultra fine grain size (< 50 nm) or by a dimensionality limited to 50 nm. Nanomaterials can be created with various modulation dimensionalities zero (atomic clusters, filaments and cluster assemblies), one (multilayers), two (ultrafine-grained overlayers or buried layers), and three (nanophase materials consisting of equiaxed nanometer sized grains) as shown in the above figure. These materials have created a high interest in recent years by virtue of their unusual mechanical, electrical, optical and magnetic properties. Some examples are given below:

- Nanophase ceramics are of particular interest because they are more ductile at elevated temperatures as compared to the coarse-grained ceramics.
- Nanostructured semiconductors are known to show various non-linear optical properties. Semiconductor Q-particles also show quantum confinement effects which may lead to special properties, like the luminescence in silicon powders and silicon germanium quantum dots as infrared optoelectronic devices. Nanostructured semiconductors are used as window layers in solar cells.
- Nanosized metallic powders have been used for the production of gas tight materials, dense parts and porous coatings. Cold welding properties combined with the ductility make them suitable for metal-metal bonding especially in the electronic industry.
- Single nanosized magnetic particles are mono-domains and one expects that also in magnetic nanophase materials the grains correspond with domains, while boundaries on the contrary to disordered walls. Very small particles have special atomic structures with discrete electronic states, which give rise to special properties in addition to the superparamagnetism behaviour. Magnetic nanocomposites have been used for mechanical force transfer (ferrofluids), for high density information storage and magnetic refrigeration.
- Nanostructured metal clusters and colloids of mono- or plurimetallic composition have a special impact in catalytic applications. They may serve as precursors for new type of heterogeneous catalysts (Cortex-catalysts) and have been shown to offer substantial advantages concerning activity, selectivity and lifetime in chemical transformations and electrocatalysis (fuel cells). Enantioselective catalysis was also achieved using chiral modifiers on the surface of nanoscale metal particles.
- Nanostructured metal-oxide thin films are receiving a growing attention for the realization of gas sensors (NOx, CO, CO_2, CH_4 and aromatic hydrocarbons) with enhanced

sensitivity and selectivity. Nanostructured metal-oxide (MnO_2) finds application for rechargeable batteries for cars or consumer goods. Nanocrystalline silicon films for highly transparent contacts in thin film solar cell and nano-structured titanium oxide porous films for its high transmission and significant surface area enhancement leading to strong absorption in dye sensitized solar cells.

- Polymer based composites with a high content of inorganic particles leading to a high dielectric constant are interesting materials for photonic band gap structure.

Examples of Nanomaterials

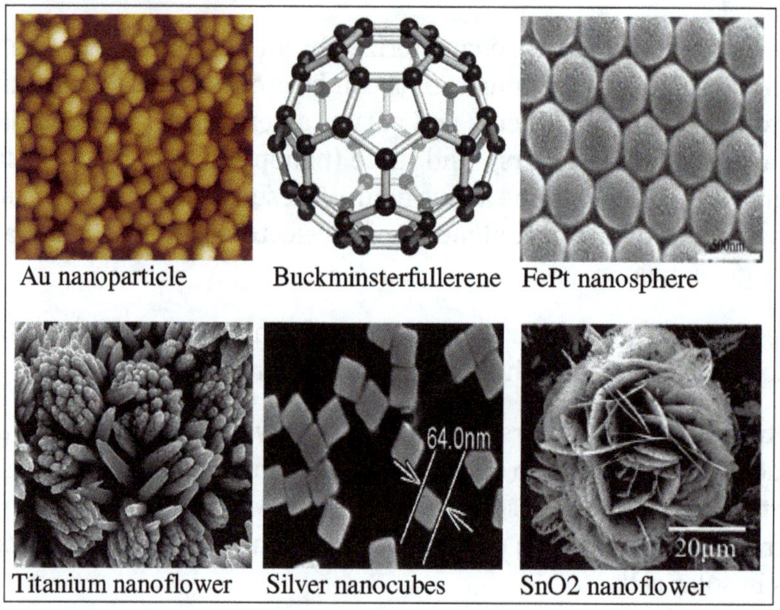

Nanomaterial Synthesis and Processing

We are dealing with very fine structures: a nanometer is a billionth of a meter. This indeed allows us to think in both the 'bottom up' or the 'top down' approaches to synthesize nanomaterials, i.e. either to assemble atoms together or to dis-assemble (break, or dissociate) bulk solids into finer pieces until they are constituted of only a few atoms. This domain is a pure example of interdisciplinary work encompassing physics, chemistry, and engineering upto medicine.

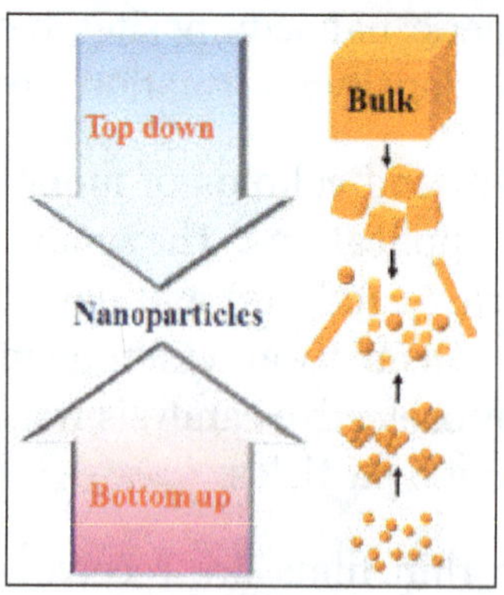

Schematic illustration of the preparative methods of nanoparticles.

Methods for Creating Nanostructures

There are many different ways of creating nanostructures: of course, macromolecules or nanoparticles or buckyballs or nanotubes and so on can be synthesized artificially for certain specific materials. They can also be arranged by methods based on equilibrium or near equilibrium thermodynamics such as methods of self-organization and self-assembly (sometimes also called bio-mimetic processes). Using these methods, synthesized materials can be arranged into useful shapes so that finally the material can be applied to a certain application.

Mechanical Grinding

Mechanical attrition is a typical example of 'top down' method of synthesis of nanomaterials, where the material is prepared not by cluster assembly but by the structural decomposition of coarser-grained structures as the result of severe plastic deformation. This has become a popular method to make nanocrystalline materials because of its simplicity, the relatively inexpensive equipment needed, and the applicability to essentially the synthesis of all classes of materials. The major advantage often quoted is the possibility for easily scaling up to tonnage quantities of material for various applications. Similarly, the serious problems that are usually cited are:

- Contamination from milling media and/or atmosphere.
- To consolidate the powder product without coarsening the nanocrystalline microstructure.

In fact, the contamination problem is often given as a reason to dismiss the method, at least for some materials. Here we will review the mechanisms presently believed responsible for formation of nanocrystalline structures by mechanical attrition of single phase powders, mechanical alloying of dissimilar powders, and mechanical crystallisation of amorphous materials. The two important problems of contamination and powder consolidation will be briefly considered.

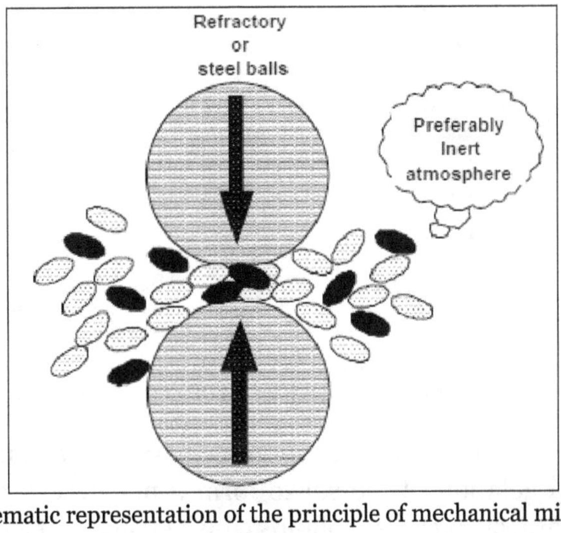

Schematic representation of the principle of mechanical milling.

Mechanical milling is typically achieved using high energy shaker, planetary ball, or tumbler mills. The energy transferred to the powder from refractory or steel balls depends on the rotational (vibrational) speed, size and number of the balls, ratio of the ball to powder mass, the time of milling and the milling atmosphere. Nanoparticles are produced by the shear action during grinding.

Milling in cryogenic liquids can greatly increase the brittleness of the powders influencing the fracture process. As with any process that produces fine particles, an adequate step to prevent oxidation is necessary. Hence this process is very restrictive for the production of non-oxide materials since then it requires that the milling take place in an inert atmosphere and that the powder particles be handled in an appropriate vacuum system or glove box. This method of synthesis is suitable for producing amorphous or nanocrystalline alloy particles, elemental or compound powders. If the mechanical milling imparts sufficient energy to the constituent powders a homogeneous alloy can be formed. Based on the energy of the milling process and thermodynamic properties of the constituents the alloy can be rendered amorphous by this processing.

Wet Chemical Synthesis of Nanomaterials

In principle we can classify the wet chemical synthesis of nanomaterials into two broad groups:

- The top down method: Where single crystals are etched in an aqueous solution for producing nanomaterials, For example, the synthesis of porous silicon by electrochemical etching.
- The bottom up method: Consisting of sol-gel method, precipitation etc. where materials containing the desired precursors are mixed in a controlled fashion to form a colloidal solution.

Sol-Gel Process

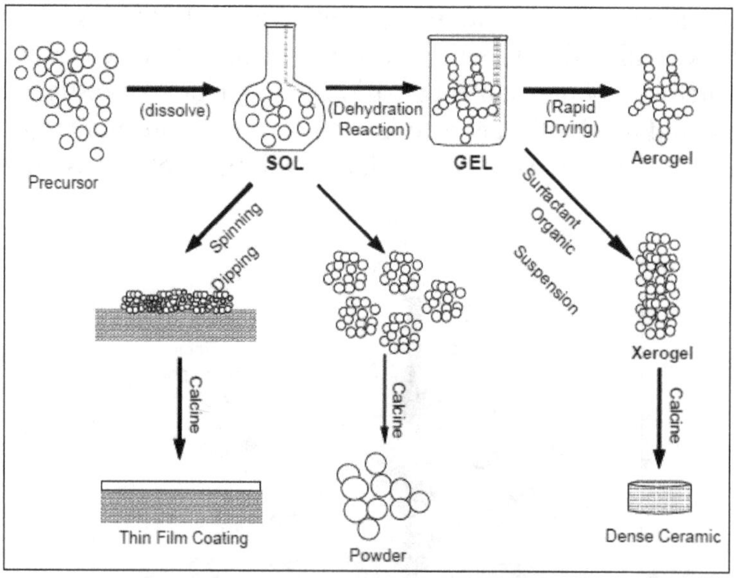

Schematic representation of sol-gel process of synthesis of nanomaterials.

The sol-gel process, involves the evolution of inorganic networks through the formation of a colloidal suspension (sol) and gelation of the sol to form a network in a continuous liquid phase (gel). The precursors for synthesizing these colloids consist usually of a metal or metalloid element surrounded by various reactive ligands. The starting material is processed to form a dispersible oxide and forms a sol in contact with water or dilute acid. Removal of the liquid from the sol yields the gel, and the sol/gel transition controls the particle size and shape. Calcination of the gel produces the oxide. Sol-gel processing refers to the hydrolysis and condensation of alkoxide-based precursors such as $Si(OEt)_4$ (tetraethyl orthosilicate, or TEOS). The reactions involved in the sol-gel chemistry based on the hydrolysis and condensation of metal alkoxides $M(OR)_z$ can be described as follows:

$$MOR + H_2O \rightarrow MOH + ROH\,(\text{hydrolysis})$$
$$MOH + ROM \rightarrow M\text{-}O\text{-}M + ROH\,(\text{condensation})$$

Sol-gel method of synthesizing nanomaterials is very popular amongst chemists and is widely employed to prepare oxide materials. The sol-gel process can be characterized by a series of distinct steps:

- Formation of different stable solutions of the alkoxide or solvated metal precursor.
- Gelation resulting from the formation of an oxide- or alcohol- bridged network (the gel) by a polycondensation reaction that results in a dramatic increase in the viscocity of the solution.
- Aging of the gel (Syneresis), during which the polycondensation reactions continue until the gel transforms into a solid mass, accompanied by contraction of the gel network and expulsion of solvent from gel pores. Ostwald ripening (also referred to as coarsening, is the phenomenon by which smaller particles are consumed by larger particles during the growth process) and phase transformations may occur concurrently with syneresis. The aging process of gels can exceed 7 days and is critical to the prevention of cracks in gels that have been cast.
- Drying of the gel, when water and other volatile liquids are removed from the gel network. This process is complicated due to fundamental changes in the structure of the gel. The drying process has itself been broken into four distinct steps:
    - The constant rate period.
    - The critical point.
    - The falling rate period.
    - The second falling rate period. If isolated by thermal evaporation, the resulting monolith is termed a xerogel. If the solvent (such as water) is extracted under supercritical or near super critical conditions, the product is an aerogel.
- Dehydration, during which surface- bound M-OH groups are removed, there by stabilizing the gel against rehydration. This is normally achieved by calcining the monolith at temperatures up to 800 °C.
- Densification and decomposition of the gels at high temperatures (T>800 °C). The pores of the gel network are collapsed, and remaining organic species are volatilized.

The interest in this synthesis method arises due to the possibility of synthesizing nonmetallic inorganic materials like glasses, glass ceramics or ceramic materials at very low temperatures compared to the high temperature process required by melting glass or firing ceramics. The major difficulties to overcome in developing a successful bottom-up approach is controlling the growth of the particles and then stopping the newly formed particles from agglomerating. Other technical issues are ensuring the reactions are complete so that no unwanted reactant is left on the product and completely removing any growth aids that may have been used in the process. Also production rates of nano powders are very low by this process. The main advantage is one can get monosized nano particles by any bottom up approach.

Gas Phase Synthesis of Nanomaterials

The gas-phase synthesis methods are of increasing interest because they allow elegant way to control process parameters in order to be able to produce size, shape and chemical composition controlled nanostructures. Before we discuss a few selected pathways for gas-phase formation of nanomaterials, some general aspects of gas-phase synthesis needs to be discussed. In conventional chemical vapour deposition (CVD) synthesis, gaseous products either are allowed to react homogeneously or heterogeneously depending on a particular application.

- In homogeneous CVD, particles form in the gas phase and diffuse towards a cold surface due to thermophoretic forces, and can either be scrapped of from the cold surface to give nano-powders, or deposited onto a substrate to yield what is called 'particulate films'.
- In heterogeneous CVD, the solid is formed on the substrate surface, which catalyses the reaction and a dense film is formed.

In order to form nanomaterials several modified CVD methods have been developed. Gas phase processes have inherent advantages, some of which are noted here:

- An excellent control of size, shape, crystallinity and chemical composition.
- Highly pure materials can be obtained.
- Multicomonent systems are relatively easy to form.
- Easy control of the reaction mechanisms.

Most of the synthesis routes are based on the production of small clusters that can aggregate to form nano particles (condensation). Condensation occurs only when the vapour is supersaturated and in these processes homogeneous nucleation in the gas phase is utilised to form particles. This can be achieved both by physical and chemical methods.

Furnace

The simplest fashion to produce nanoparticles is by heating the desired material in a heat resistant crucible containing the desired material. This method is appropriate only for materials that have a high vapour pressure at the heated temperatures that can be as high as 2000 °C. Energy is normally introduced into the precursor by arc heating, electron-beam heating or Joule heating. The atoms are evaporated into an atmosphere, which is either inert (e.g. He) or reactive (so as to form a compound). To carry out reactive synthesis, materials with very low vapour pressure have to be fed into the furnace in the form of a suitable precursor such as organometallics, which decompose in the furnace to produce a condensable material. The hot atoms of the evaporated matter lose energy by collision with the atoms of the cold gas and undergo condensation into small clusters via homogeneous nucleation. In case a compound is being synthesized, these precursors react in the gas phase and form a compound with the material that is separately injected in the reaction chamber. The clusters would continue to grow if they remain in the supersaturated region. To control their size, they need to be rapidly removed from the supersaturated environment by a carrier gas. The cluster size and its distribution are controlled by only three parameters:

- The rate of evaporation (energy input).
- The rate of condensation (energy removal).
- The rate of gas flow (cluster removal).

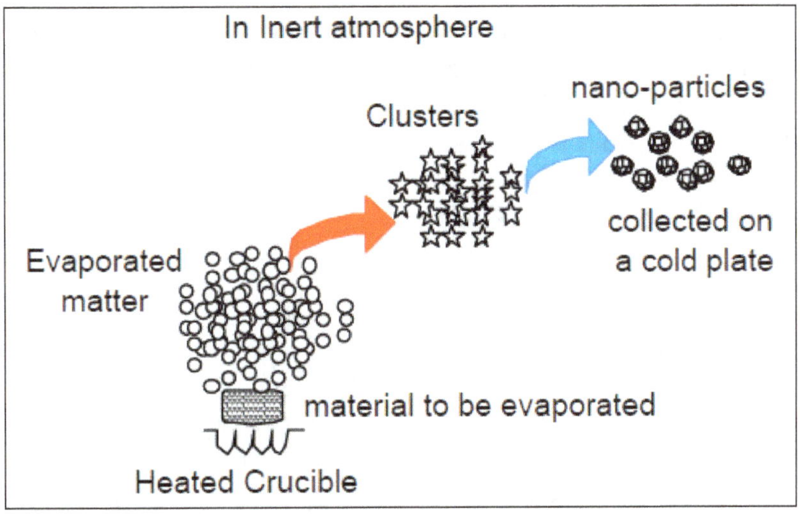

Schematic representation of gas phase process of synthesis of single phase nanomaterials from a heated crucible.

Because of its inherent simplicity, it is possible to scale up this process from laboratory (mg/day) to industrial scales (tons/day).

Flame Assisted Ultrasonic Spray Pyrolysis

In this process, precursors are nebulized and then unwanted components are burnt in a flame to get the required material, e.g. ZrO_2 has been obtained by this method from a precursor of Zr $(CH_3\ CH_2\ CH_2O)_4$. Flame hydrolysis that is a variant of this process is used for the manufacture of fused silica. In the process, silicon tetrachloride is heated in an oxy-hydrogen flame to give highly dispersed silica. The resulting white amorphous powder consists of spherical particles with sizes in the range 7-40 nm. The combustion flame synthesis, in which the burning of a gas mixture, e.g. acetylene and oxygen or hydrogen and oxygen, supplies the energy to initiate the pyrolysis of precursor compounds, is widely used for the industrial production of powders in large quantities, such as carbon black, fumed silica and titanium dioxide. However, since the gas pressure during the reaction is high, highly agglomerated powders are produced which is disadvantageous for subsequent processing. The basic idea of low pressure combustion flame synthesis is to extend the pressure range to the pressures used in gas phase synthesis and thus to reduce or avoid the agglomeration. Low pressure flames have been extensively used by aerosol scientists to study particle formation in the flame.

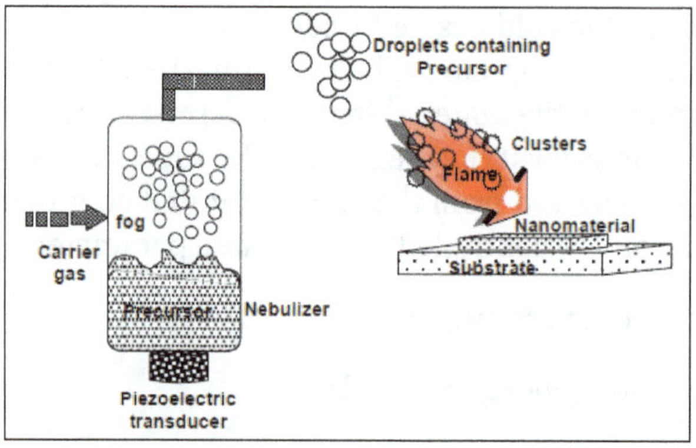

Flame assisted ultrasonic spray pyrolysis.

A key for the formation of nanoparticles with narrow size distributions is the exact control of the flame in order to obtain a flat flame front. Under these conditions the thermal history, i.e. time and temperature, of each particle formed is identical and narrow distributions result. However, due to the oxidative atmosphere in the flame, this synthesis process is limited to the formation of oxides in the reactor zone.

Gas Condensation Processing (GPC)

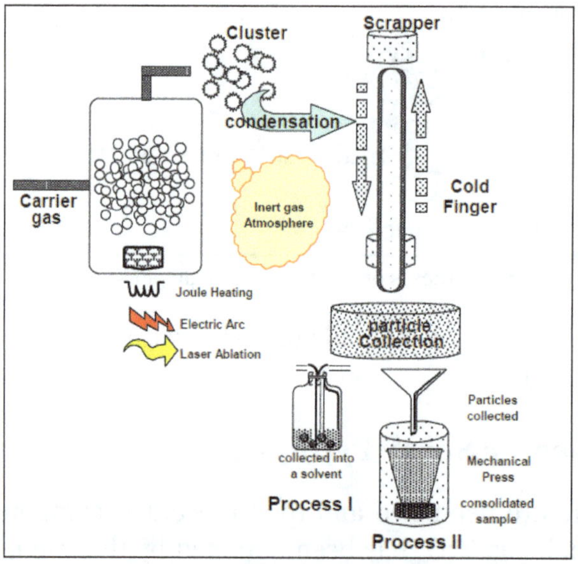

Schematic representation of typical set-up for gas condensation synthesis of nanomaterials followed by consolidation in a mechanical press or collection in an appropriate solvent media.

In this technique, a metallic or inorganic material, e.g. a suboxide, is vaporised using thermal evaporation sources such as crucibles, electron beam evaporation devices or sputtering sources in an atmosphere of 1-50 mbar He (or another inert gas like Ar, Ne, Kr). Cluster form in the vicinity of the source by homogenous nucleation in the gas phase and grow by coalescence and incorporation of atoms from the gas phase. The cluster or particle size depends critically on the residence time of the particles in the growth system and can be influenced by the gas pressure, the kind of inert gas, i.e. He, Ar or Kr, and on the evaporation rate/vapour pressure of the evaporating material. With increasing gas pressure, vapour pressure and mass of the inert gas used the average particle size of the nanoparticles increases. Lognormal size distributions have been found experimentally and have been explained theoretically by the growth mechanisms of the particles. Even in more complex processes such as the low pressure combustion flame synthesis where a number of chemical reactions are involved the size distributions are determined to be lognormal.

Originally, a rotating cylindrical device cooled with liquid nitrogen was employed for the particle collection: the nanoparticles in the size range from 2-50 nm are extracted from the gas flow by thermophoretic forces and deposited loosely on the surface of the collection device as a powder of low density and no agglomeration. Subsequenly, the nanoparticles are removed from the surface of the cylinder by means of a scraper in the form of a metallic plate. In addition to this cold finger device several techniques known from aerosol science have now been implemented for the use in gas condensation systems such as corona discharge, etc.

These methods allow for the continuous operation of the collection device and are better suited for larger scale synthesis of nanopowders. However, these methods can only be used in a system designed for gas flow, i.e. a dynamic vacuum is generated by means of both continuous pumping and gas inlet via mass flow controller. A major advantage over convectional gas flow is the improved control of the particle sizes. It has been found that the particle size distributions in gas flow systems, which are also lognormal, are shifted towards smaller average values with an appreciable reduction of the standard deviation of the distribution. Depending on the flow rate of the He-gas, particle sizes are reduced by 80% and standard deviations by 18%.

The synthesis of nanocrystalline pure metals is relatively straightforward as long as evaporation can be done from refractory metal crucibles (W, Ta or Mo). If metals with high melting points or metals which react with the crucibles, are to be prepared, sputtering, i.e. for W and Zr, or laser or electron beam evaporation has to be used. Synthesis of alloys or intermetallic compounds by thermal evaporation can only be done in the exceptional cases that the vapour pressures of the elements are similar. As an alternative, sputtering from an alloy or mixed target can be employed. Composite materials such as Cu/Bi or W/Ga have been synthesised by simultaneous evaporation from two separate crucibles onto a rotating collection device. It has been found that excellent intermixing on the scale of the particle size can be obtained.

However, control of the composition of the elements has been difficult and reproducibility is poor. Nanocrystalline oxide powders are formed by controlled postoxidation of primary nanoparticles of a pure metal (e.g. Ti to TiO_2) or a suboxide (e.g. ZrO to ZrO_2). Although the gas condensation method including the variations have been widely employed to prepared a variety of metallic and ceramic materials, quantities have so far been limited to a laboratory scale. The quantities of metals are below 1 g/day, while quantities of oxides can be as high as 20 g/day for simple oxides such as CeO_2 or ZrO_2. These quantities are sufficient for materials testing but not for industrial production. However, it should be mentioned that the scale-up of the gas condensation method for industrial production of nanocrystalline oxides by a company called nanophase technologies has been successful.

Chemical Vapour Condensation (CVC)

As shown schematically in figure, the evaporative source used in GPC is replaced by a hot wall reactor in the Chemical Vapour Condensation or the CVC process. Depending on the processing parameters nucleation of nanoparticles is observed during chemical vapour deposition (CVC) of thin films and poses a major problem in obtaining good film qualities. The original idea of the novel CVC process which is schematically shown below where, it was intended to adjust the parameter field during the synthesis in order to suppress film formation and enhance homogeneous nucleation of particles in the gas flow. It is readily found that the residence time of the precursor in the reactor determines if films or particles are formed. In a certain range of residence time both particle and film formation can be obtained.

Adjusting the residence time of the precursor molecules by changing the gas flow rate, the pressure difference between the precursor delivery system and the main chamber occurs. Then the temperature of the hot wall reactor results in the fertile production of nanosized particles of metals and ceramics instead of thin films as in CVD processing. In the simplest form a metal organic precursor is introduced into the hot zone of the reactor using mass flow controller. Besides the increased quantities in this continuous process compared to GPC has been demonstrated that a wider range of ceramics including nitrides and carbides can be synthesised. Additionally, more complex oxides such as $BaTiO_3$ or composite structures can be formed as well. The extension to production of nanoparticles requires the determination of a modified parameter field in order to promote particle formation instead of film formation. In addition to the formation of single phase nanoparticles by CVC of a single precursor the reactor allows the synthesis of:

- Mixtures of nanoparticles of two phases or doped nanoparticles by supplying two precursors at the front end of the reactor.
- Coated nanoparticles, i.e., n-ZrO_2 coated with n-Al_2O_3 or vice versa, by supplying a second precursor at a second stage of the reactor. In this case nanoparticles which have been formed by homogeneous nucleation are coated by heterogeneous nucleation in a second stage of the reactor.

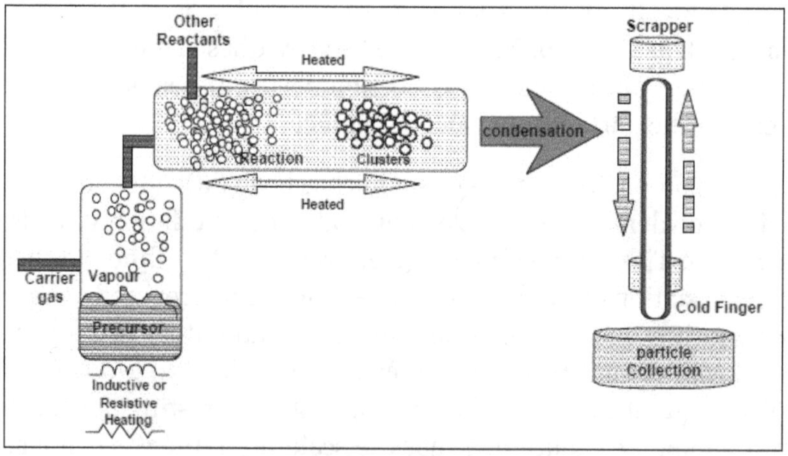

A schematic of a typical CVC reactor.

Because CVC processing is continuous, the production capabilities are much larger than in GPC processing. Quantities in excess of 20 g/hr have been readily produced with a small scale laboratory reactor. A further expansion can be envisaged by simply enlarging the diameter of the hot wall reactor and the mass flow through the reactor.

Sputtered Plasma Processing

In this method is yet again a variation of the gas-condensation method excepting the fact that the source material is a sputtering target and this target is sputtered using rare gases and the constituents are allowed to agglomerate to produce nanomaterial. Both dc (direct current) and rf (radio-frequency) sputtering has been used to synthesize nanoparticles. Again reactive sputtering or multitarget sputtering has been used to make alloys and/or oxides, carbides, nitrides of materials. This method is specifically suitable for the preparation of ultrapure and non-agglomerated nanoparticles of metal.

Microwave Plasma Processing

This technique is similar to the CVC method but employs plasma instead of high temperature for decomposition of the metal organic precursors. The method uses microwave plasma in a 50 mm diameter reaction vessel made of quartz placed in a cavity connected to a microwave generator. A precursor such as a chloride compound is introduced into the front end of the reactor. Generally, the microwave cavity is designed as a single mode cavity using the TE10 mode in a WR975 waveguide with a frequency of 0.915 GHz. The major advantage of the plasma assisted pyrolysis in contrast to the thermal activation is the low temperature reaction which reduces the tendency for agglomeration of the primary particles. This is also true in the case of plasma-CVD processes. Additionally, it has been shown that by introducing another precursor into a second reaction zone of the tubular reactor, e.g. by splitting the microwave guide tubes, the primary particles can be coated with a second phase. For example, it has been demonstrated that ZrO_2 nanoparticles can be coated by Al_2O_3. In this case the inner ZrO_2 core is crystalline, while the Al_2O_3 coating is amorphous. The reaction sequence can be reversed with the result that an amorphous Al_2O_3 core is coated with crystalline ZrO_2. While the formation of the primary particles occurs by homogeneous nucleation, it can be easily estimated using gas reaction kinetics that the coating on the primary particles grows heterogeneously and that homogeneous nucleation of nanoparticles originating from the second compound has a very low probability.

Particle Precipitation Aided CVD

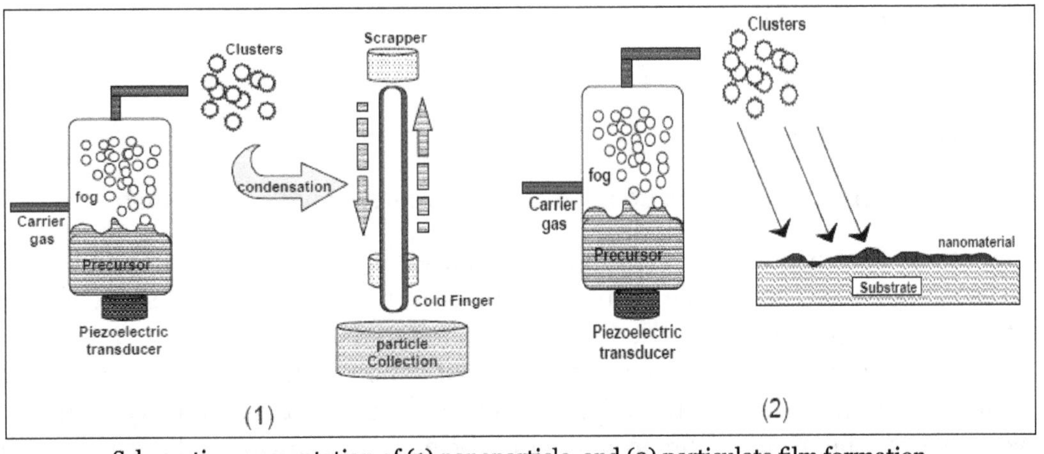

Schematic representation of (1) nanoparticle, and (2) particulate film formation.

In another variation of this process, colloidal clusters of materials are used to prepare nanoparticles. The CVD reaction conditions are so set that particles form by condensation in the gas phase and collect onto a substrate, which is kept under a different condition that allows heterogeneous nucleation. By this method both nanoparticles and particulate films can be prepared. An example of this method has been used to form nanomaterials e.g. SnO_2, by a method called pyrosol deposition process, where clusters of tin hydroxide are transformed into small aerosol droplets, following which they are reacted onto a heated glass substrate.

Laser Ablation

Laser ablation has been extensively used for the preparation of nanoparticles and particulate films. In this process a laser beam is used as the primary excitation source of ablation for generating clusters directly from a solid sample in a wide variety of applications. The small dimensions of the particles and the possibility to form thick films make this method quite an efficient tool for the production of ceramic particles and coatings and also an ablation source for analytical applications such as the coupling to induced coupled plasma emission spectrometry, ICP, the formation of the nanoparticles has been explained following a liquefaction process which generates an aerosol, followed by the cooling/solidification of the droplets which results in the formation of fog. The general dynamics of both the aerosol and the fog favors the aggregation process and micrometer-sized fractal-like particles are formed. The laser spark atomizer can be used to produce highly mesoporous thick films and the porosity can be modified by the carrier gas flow rate. ZrO_2 and SnO_2 nanoparticulate thick films were also synthesized successfully using this process with quite identical microstructure. Synthesis of other materials such as lithium manganate, silicon and carbon has also been carried out by this technique.

Properties of Nanomaterials

Nanomaterials have the structural features in between of those of atoms and the bulk materials. While most microstructured materials have similar properties to the corresponding bulk materials, the properties of materials with nanometer dimensions are significantly different from those of atoms and bulks materials. This is mainly due to the nanometer size of the materials which render them:

- Large fraction of surface atoms.
- High surface energy.
- Spatial confinement.
- Reduced imperfections, which do not exist in the corresponding bulk materials.

Due to their small dimensions, nanomaterials have extremely large surface area to volume ratio, which makes a large to be the surface or interfacial atoms, resulting in more "surface" dependent material properties. Especially when the sizes of nanomaterials are comparable to length, the entire material will be affected by the surface properties of nanomaterials. This in turn may enhance or modify the properties of the bulk materials. For example, metallic nanoparticles can be used as very active catalysts. Chemical sensors from nanoparticles and nanowires enhanced the sensitivity and sensor selectivity. The nanometer feature sizes of nanomaterials also have spatial confinement effect on the materials, which bring the quantum effects.

The energy band structure and charge carrier density in the materials can be modified quite differently from their bulk and in turn will modify the electronic and optical properties of the materials.

For example, lasers and light emitting diodes (LED) from both of the quantum dots and quantum wires are very promising in the future optoelections. High density information storage using quantum dot devices is also a fast developing area. Reduced imperfections are also an important factor in determination of the properties of the nanomaterials. Nanosturctures and Nanomaterials favors of a self-purification process in that the impurities and intrinsic material defects will move to near the surface upon thermal annealing. This increased materials perfection affects the properties of nanomaterials. For example, the chemical stability for certain nanomaterials may be enhanced, the mechanical properties of nanomaterials will be better than the bulk materials. The superior mechanical properties of carbon nanotubes are well known. Due to their nanometer size, nanomaterials are already known to have many novel properties. Many novel applications of the nanomaterials rose from these novel properties have also been proposed.

Optical Properties

One of the most fascinating and useful aspects of nanomaterials is their optical properties. Applications based on optical properties of nanomaterials include optical detector, laser, sensor, imaging, phosphor, display, solar cell, photocatalysis, photoelectrochemistry and biomedicine.

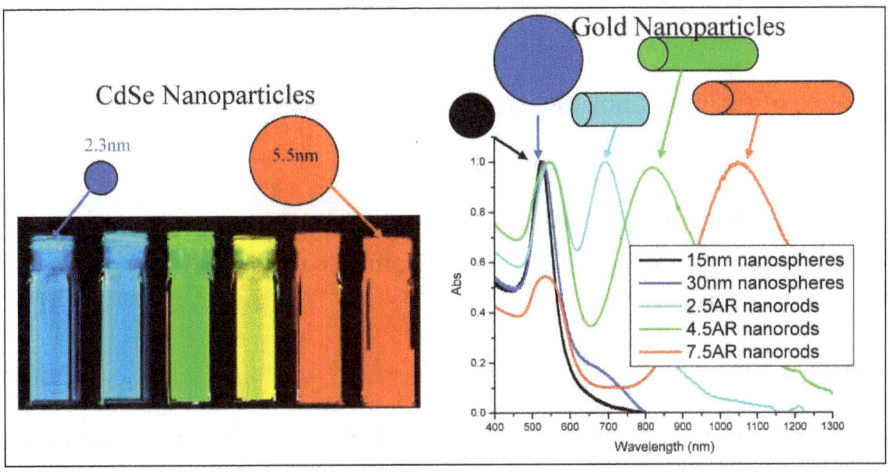

Fluorescence emission of (CdSe) ZnS quantum dots of various sizes and absorption spectra of various sizes and shapes of gold nanoparticles.

The optical properties of nanomaterials depend on parameters such as feature size, shape, surface characteristics, and other variables including doping and interaction with the surrounding environment or other nanostructures. Likewise, shape can have dramatic influence on optical properties of metal nanostructures. Figure shown above exemplifies the difference in the optical properties of metal and semiconductor nanoparticles. With the CdSe semiconductor nanoparticles, a simple change in size alters the optical properties of the nanoparticles. When metal nanoparticles are enlarged, their optical properties change only slightly as observed for the different samples of gold nanospheres. However, when an anisotropy is added to the nanoparticle, such as growth of nanorods, the optical properties of the nanoparticles change dramatically.

Electrical Properties

Electrical Properties of Nanoparticles discuss about fundamentals of electrical conductivity in nanotubes and nanorods, carbon nanotubes, photoconductivity of nanorods, electrical conductivity of nanocomposites. One interesting method which can be used to demonstrate the steps in conductance is the mechanical thinning of a nanowire and measurement of the electrical current at a constant applied voltage. The important point here is that, with decreasing diameter of the wire, the number of electron wave modes contributing to the electrical conductivity is becoming increasingly smaller by well-defined quantized steps.

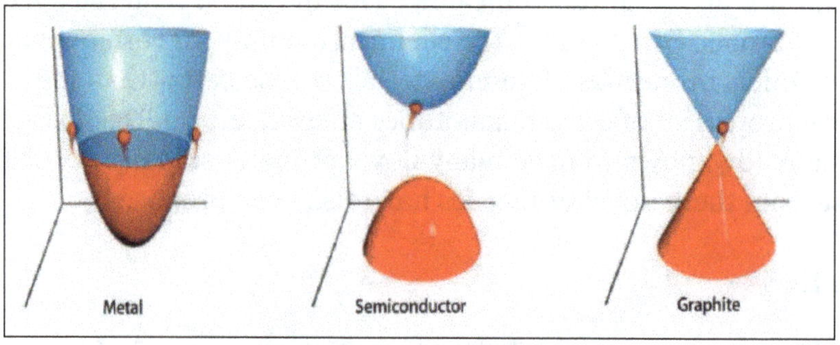

Electrical behavior of naotubes.

In electrically conducting carbon nanotubes, only one electron wave mode is observed which transport the electrical current. As the lengths and orientations of the carbon nanotubes are different, they touch the surface of the mercury at different times, which provides two sets of information:

- The influence of carbon nanotube length on the resistance.
- The resistances of the different nanotubes.

As the nanotubes have different lengths, then with increasing protrusion of the fiber bundle an increasing number of carbon nanotubes will touch the surface of the mercury droplet and contribute to the electrical current transport.

Mechanical Properties

"Mechanical Properties of Nanoparticles" deals with bulk metallic and ceramic materials, influence of porosity, influence of grain size, superplasticity, filled polymer composites, particle filled polymers, polymer-based nanocomposites filled with platelets, carbon nanotube-based composites. The discussion of mechanical properties of nanomaterials is, in to some extent, only of quite basic interest, the reason being that it is problematic to produce macroscopic bodies with a high density and a grain size in the range of less than 100 nm. However, two materials, neither of which is produced by pressing and sintering, have attracted much greater interest as they will undoubtedly achieve industrial importance.

These materials are polymers which contain nanoparticles or nanotubes to improve their mechanical behaviors, and severely plastic-deformed metals, which exhibit astonishing properties. However, because of their larger grain size, the latter are generally not accepted as nanomaterials. Experimental studies on the mechanical properties of bulk nanomaterials are generally impaired by major experimental problems in producing specimens with exactly defined grain sizes and porosities. Therefore, model calculations and molecular dynamic studies are of major importance for an understanding of the mechanical properties of these materials.

Filling polymers with nanoparticles or nanorods and nanotubes, respectively, leads to significant improvements in their mechanical properties. Such improvements depend heavily on the type of the filler and the way in which the filling is conducted. The latter point is of special importance, as any specific advantages of a nanoparticulate filler may be lost if the filler forms aggregates, thereby mimicking the large particles. Particulate-filled polymer-based nanocomposites exhibit a broad range of failure strengths and strains. This depends on the shape of the filler, particles or platelets, and on the degree of agglomeration. In this class of material, polymers filled with silicate platelets exhibit the best mechanical properties and are of the greatest economic relevance. The larger the particles of the filler or agglomerates, the poorer are the properties obtained.

Although, potentially, the best composites are those filled with nanofibers or nanotubes, experience teaches that sometimes such composites have the least ductility. On the other hand, by using carbon nanotubes it is possible to produce composite fibers with extremely high strength and strain at rupture. Among the most exciting nanocomposites are the polymerceramic nanocomposites, where the ceramic phase is platelet-shaped. This type of composite is preferred in nature, and is found in the structure of bones, where it consists of crystallized mineral platelets of a few nanometers thickness that are bound together with collagen as the matrix. Composites consisting of a polymer matrix and defoliated phyllosilicates exhibit excellent mechanical and thermal properties.

Magnetic Properties

Bulk gold and Pt are non-magnetic, but at the nano size they are magnetic. Surface atoms are not only different to bulk atoms, but they can also be modified by interaction with other chemical species, that is, by capping the nanoparticles. This phenomenon opens the possibility to modify the physical properties of the nanoparticles by capping them with appropriate molecules. Actually, it should be possible that non-ferromagnetic bulk materials exhibit ferromagnetic-like behavior when prepared in nano range. One can obtain magnetic nanoparticles of Pd, Pt and the surprising case of Au (that is diamagnetic in bulk) from non-magnetic bulk materials. In the case of Pt and Pd, the ferromagnetism arises from the structural changes associated with size effects.

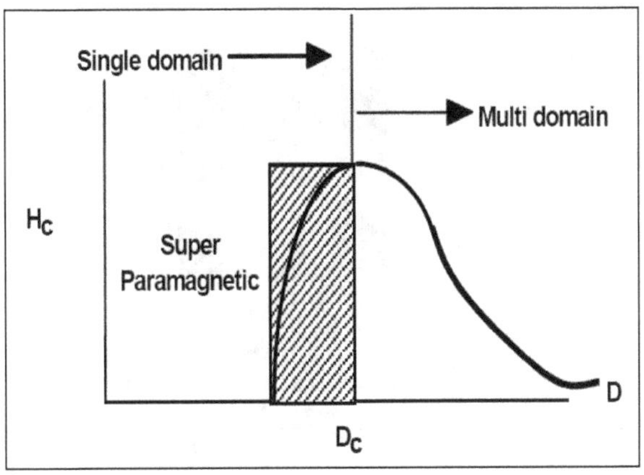

Magnetic properties of nanostrucutred materials.

However, gold nanoparticles become ferromagnetic when they are capped with appropriate molecules: the charge localized at the particle surface gives rise to ferromagnetic-like behavior. Surface and the core of Au nanoparticles with 2 nm in diameter show ferromagnetic and paramagnetic character, respectively. The large spin-orbit coupling of these noble metals can yield to a large anisotropy and therefore exhibit high ordering temperatures. More surprisingly, permanent magnetism was observed up to room temperature for thiol-capped Au nanoparticles. For nanoparticles with sizes below 2 nm the localized carriers are in the 5d band. Bulk Au has an extremely low density of states and becomes diamagnetic, as is also the case for bare Au nanoparticles. This observation suggested that modification of the d band structure by chemical bonding can induce ferromagnetic like character in metallic clusters.

Selected Application of Nanomaterials

Nanomaterials having wide range of applications in the field of electronics, fuel cells, batteries, agriculture, food industry and medicines, etc. It is evident that nanomaterials split their conventional counterparts because of their superior chemical, physical, and mechanical properties and of their exceptional formability.

Fuel Cells

A fuel cell is an electrochemical energy conversion device that converts the chemical energy from fuel (on the anode side) and oxidant (on the cathode side) directly into electricity. The heart of fuel cell is the electrodes. The performance of a fuel cell electrode can be optimized in two ways; by improving the physical structure and by using more active electro catalyst. A good structure of electrode must provide ample surface area, provide maximum contact of catalyst, reactant gas and electrolyte, facilitate gas transport and provide good electronic conductance. In this fashion the structure should be able to minimize losses.

Carbon Nanotubes - Microbial Fuel Cell

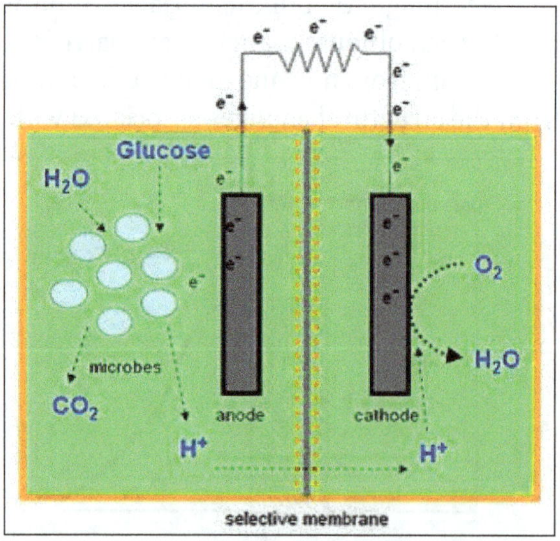

Schematic representation of microbial fuel cell.

Microbial fuel cell is a device in which bacteria consume water-soluble waste such as sugar, starch and alcohols and produces electricity plus clean water. This technology will make it possible to generate electricity while treating domestic or industrial wastewater. Microbial fuel cell can turn different carbohydrates and complex substrates present in wastewaters into a source of electricity. The efficient electron transfer between the microorganism and the anode of the microbial fuel cell plays a major role in the performance of the fuel cell. The organic molecules present in the wastewater posses a certain amount of chemical energy, which is released when converting them to simpler molecules like CO_2. The microbial fuel cell is thus a device that converts the chemical energy present in water-soluble waste into electrical energy by the catalytic reaction of microorganisms.

Carbon nanotubes (CNTs) have chemical stability, good mechanical properties and high surface area, making them ideal for the design of sensors and provide very high surface area due to its structural network. Since carbon nanotubes are also suitable supports for cell growth, electrodes of microbial fuel cells can be built using of CNT. Due to three-dimensional architectures and enlarged electrode surface area for the entry of growth medium, bacteria can grow and proliferate and get immobilized. Multi walled CNT scaffolds could offer self-supported structure with large surface area through which hydrogen producing bacteria (e.g., E. coli) can eventually grow and proliferate. Also CNTs and MWCNTs have been reported to be biocompatible for different eukaryotic cells. The efficient proliferation of hydrogen producing bacteria throughout an electron conducting scaffold of CNT can form the basis for the potential application as electrodes in MFCs leading to efficient performance.

Catalysis

Higher surface area available with the nanomaterial counterparts, nano-catalysts tend to have exceptional surface activity. For example, reaction rate at nano-aluminum can go so high, that it is utilized as a solid-fuel in rocket propulsion, whereas the bulk aluminum is widely used in utensils. Nano-aluminum becomes highly reactive and supplies the required thrust to send off pay loads in space. Similarly, catalysts assisting or retarding the reaction rates are dependent on the surface activity, and can very well be utilized in manipulating the rate-controlling step.

Phosphors for High-Definition TV

The resolution of a television, or a monitor, depends greatly on the size of the pixel. These pixels are essentially made of materials called “phosphors,” which glow when struck by a stream of electrons inside the cathode ray tube (CRT). The resolution improves with a reduction in the size of the pixel, or the phosphors. Nanocrystalline zinc selenide, zinc sulfide, cadmium sulfide, and lead telluride synthesized by the sol-gel techniques are candidates for improving the resolution of monitors. The use of nanophosphors is envisioned to reduce the cost of these displays so as to render high-definition televisions (HDTVs) and personal computers affordable to be purchase.

Next-Generation Computer Chips

The microelectronics industry has been emphasizing miniaturization, whereby the circuits, such as transistors, resistors, and capacitors, are reduced in size. By achieving a significant reduction in their size, the microprocessors, which contain these components, can run much faster, thereby enabling computations at far greater speeds. However, there are several technological impediments to these advancements, including lack of the ultrafine precursors to manufacture these components; poor dissipation of tremendous amount of heat generated by these microprocessors due to faster speeds; short mean time to failures (poor reliability), etc. Nanomaterials help the industry break these barriers down by providing the manufacturers with nanocrystalline starting materials, ultra-high purity materials, materials with better thermal conductivity, and longer-lasting, durable interconnections (connections between various components in the microprocessors).

For example: Nanowires for junctionless transistors: Transistors are made so tiny to reduce the size of sub-assemblies of electronic systems and make smaller and smaller devices, but it is difficult to create high-quality junctions. In particular, it is very difficult to change the doping concentration of a material over distances shorter than about 10 nm. Researchers have succeeded in making the junctionless transistor having nearly ideal electrical properties. It could potentially operate faster and use less power than any conventional transistor on the market today. The device consists of a silicon nanowire in which current flow is perfectly controlled by a silicon gate that is separated from the nanowire by a thin insulating layer. The entire silicon nanowire is heavily n-doped, making it an excellent conductor. However, the gate is p-doped and its presence has the effect of depleting the number of electrons in the region of the nanowire under the gate. The device also has near-ideal electrical properties and behaves like the most perfect of transistors without suffering from current leakage like conventional devices and operates faster and using less energy.

Silicon nanowires in junctionless transistors.

Elimination of Pollutants

Nanomaterials possess extremely large grain boundaries relative to their grain size. Hence, they are very active in terms of their chemical, physical, and mechanical properties. Due to their enhanced chemical activity, nanomaterials can be used as catalysts to react with such noxious and toxic gases as carbon monoxide and nitrogen oxide in automobile catalytic converters and power generation equipment to prevent environmental pollution arising from burning gasoline and coal.

Sun-Screen Lotion

Prolonged UV exposure causes skin-burns and cancer. Sun-screen lotions containing nano-TiO_2 provide enhanced sun protection factor (SPF) while eliminating stickiness. The added advantage of nano skin blocks (ZnO and TiO_2) arises as they protect the skin by sitting onto it rather than penetrating into the skin. Thus they block UV radiation effectively for prolonged duration. Additionally, they are transparent, thus retain natural skin color while working better than conventional skin-lotions.

Sensors

Sensors rely on the highly active surface to initiate a response with minute change in the concentration of the species to be detected. Engineered monolayers (few Angstroms thick) on the sensor surface are exposed to the environment and the peculiar functionality (such as change in potential as the CO/anthrax level is detected) is utilized in sensing.

Disadvantages of Nanomaterials

- Instability of the particles: Retaining the active metal nanoparticles is highly challenging, as the kinetics associated with nanomaterials is rapid. In order to retain nanosize of particles, they are encapsulated in some other matrix. Nanomaterials are thermodynamically metastable and lie in the region of high-energy local-minima. Hence they are prone to attack and undergo transformation. These include poor corrosion resistance, high solubility, and phase change of nanomaterials. This leads to deterioration in properties and retaining the structure becomes challenging. Fine metal particles act as strong explosives owing to their high surface area coming in direct contact with oxygen. Their exothermic combustion can easily cause explosion.
- Impurity: Because nanoparticles are highly reactive, they inherently interact with impurities as well. In addition, encapsulation of nanoparticles becomes necessary when they are synthesized in a solution (chemical route). The stabilization of nanoparticles occurs because of a non-reactive species engulfing the reactive nano-entities. Thereby, these secondary impurities become a part of the synthesized nanoparticles, and synthesis of pure nanoparticles becomes highly difficult. Formation of oxides, nitrides, etc. can also get aggravated from the impure environment/ surrounding while synthesizing nanoparticles. Hence retaining high purity in nanoparticles can become a challenge hard to overcome.
- Biologically harmful: Nanomaterials are usually considered harmful as they become transparent to the cell-dermis. Toxicity of nanomaterials also appears predominant owing to their high surface area and enhanced surface activity. Nanomaterials have shown to cause irritation, and have indicated to be carcinogenic. If inhaled, their low mass entraps them inside lungs, and in no way they can be expelled out of body. Their interaction with liver/blood could also prove to be harmful (though this aspect is still being debated on).
- Difficulty in synthesis, isolation and application: It is extremely hard to retain the size of nanoparticles once they are synthesized in a solution. Hence, the nanomaterials have to be encapsulated in a bigger and stable molecule/material. Hence free nanoparticles are hard to be utilized in isolation, and they have to be interacted for intended use via secondary means of exposure. Grain growth is inherently present in nanomateirals during their processing. The finer grains tend to merge and become bigger and stable grains at high temperatures and times of processing.
- Recycling and disposal: There are no hard-and-fast safe disposal policies evolved for nanomaterials. Issues of their toxicity are still under question, and results of exposure experiments are not available. Hence the uncertainty associated with affects of nanomaterials is yet to be assessed in order to develop their disposal policies.

Nanoparticle

Nanoparticle is an ultrafine unit with dimensions measured in nanometres (nm; 1 nm = 10^{-9} metre). Nanoparticles exist in the natural world and are also created as a result of human activities. Because of their submicroscopic size, they have unique material characteristics, and manufactured nanoparticles may find practical applications in a variety of areas, including medicine, engineering, catalysis, and environmental remediation.

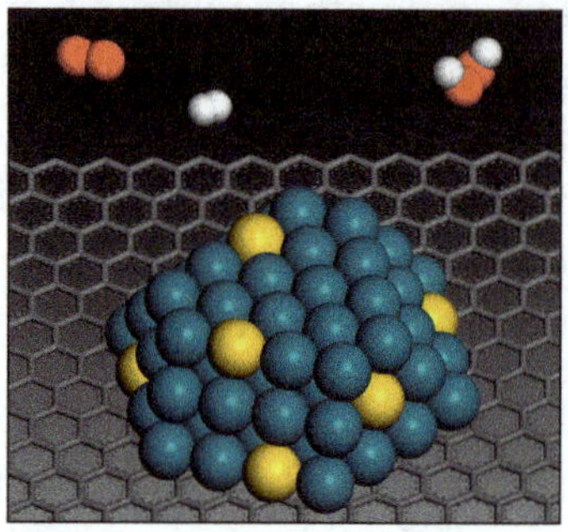

Image showing nanoparticles of an alloy of gold (yellow) and palladium (blue) on an acid-treated carbon support (gray). These particles were employed as catalysts for the formation of hydrogen peroxide from hydrogen (white) and oxygen (red).

Properties of Nanoparticles

In 2008 the International Organization for Standardization (ISO) defined a nanoparticle as a discrete nano-object where all three Cartesian dimensions are less than 100 nm. The ISO standard similarly defined two-dimensional nano-objects (i.e., nanodiscs and nanoplates) and one-dimensional nano-objects (i.e., nanofibres and nanotubes). But in 2011 the Commission of the European Union endorsed a more-technical but wider-ranging definition.

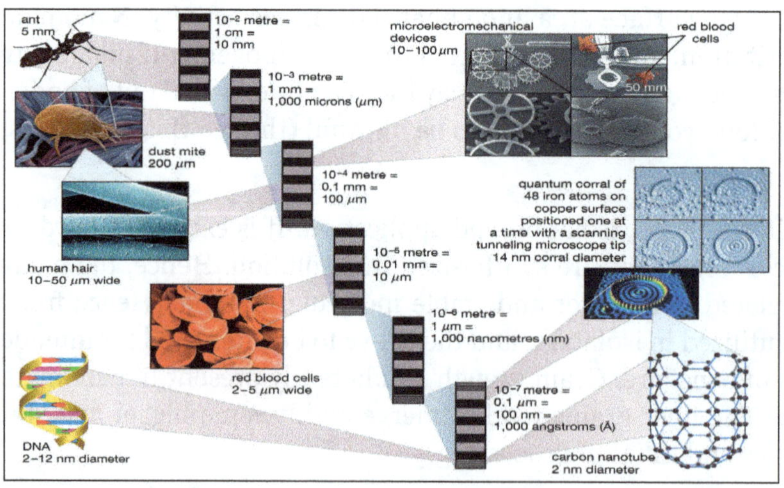

Examples from biological and mechanical realms illustrate various "orders of magnitude" (powers of 10), from 10^{-2} metre down to 10^{-7} metre.

A natural, incidental or manufactured material containing particles, in an unbound state or as an aggregate or as an agglomerate and where, for 50% or more of the particles in the number size distribution, one or more external dimensions is in the size range 1 nm–100 nm. Under that definition a nano-object needs only one of its characteristic dimensions to be in the range 1–100 nm to be classed as a nanoparticle, even if its other dimensions are outside that range. (The lower limit of 1 nm is used because atomic bond lengths are reached at 0.1 nm.) That size range from 1 to 100 nm overlaps considerably with that previously assigned to the field of colloid science from 1 to 1,000 nm which is sometimes alternatively called the mesoscale. Thus, it is not uncommon to find literature that refers to nanoparticles and colloidal particles in equal terms. The difference is essentially semantic for particles below 100 nm in size.

Nanoparticles can be classified into any of various types, according to their size, shape, and material properties. Some classifications distinguish between organic and inorganic nanoparticles; the first group includes dendrimers, liposomes, and polymeric nanoparticles, while the latter includes fullerenes, quantum dots, and gold nanoparticles. Other classifications divide nanoparticles according to whether they are carbon-based, ceramic, semiconducting, or polymeric. In addition, nanoparticles can be classified as hard (e.g., titania [titanium dioxide], silica [silica dioxide] particles, and fullerenes) or as soft (e.g., liposomes, vesicles, and nanodroplets). The way in which nanoparticles are classified typically depends on their application, such as in diagnosis or therapy versus basic research, or may be related to the way in which they were produced.

There are three major physical properties of nanoparticles, and all are interrelated:

- They are highly mobile in the free state (e.g., in the absence of some other additional influence, a 10-nm-diameter nanosphere of silica has a sedimentation rate under gravity of 0.01 mm/day in water).
- They have enormous specific surface areas (e.g., a standard teaspoon, or about 6 ml, of 10-nm-diameter silica nanospheres has more surface area than a dozen doubles-sized tennis courts; 20 percent of all the atoms in each nanosphere will be located at the surface).
- They may exhibit what are known as quantum effects. Thus, nanoparticles have a vast range of compositions, depending on the use or the product.

Nanoparticle-Based Technologies

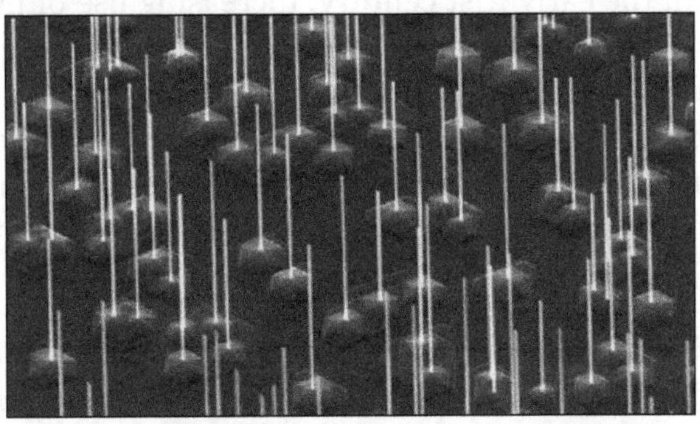

Nanowires as seen by a field-emission microscope.

In general, nanoparticle-based technologies centre on opportunities for improving the efficiency, sustainability, and speed of already-existing processes. That is possible because, relative to the materials used traditionally for industrial processes (e.g., industrial catalysis), nanoparticle-based technologies use less material, a large proportion of which is already in a more "reactive" state.

Other opportunities for nanoparticle-based technologies include the use of nanoscale zero-valent iron (NZVI) particles as a field-deployable means of remediating organochlorine compounds, such as polychlorinated biphenyls (PCBs), in the environment. NZVI particles are able to permeate into rock layers in the ground and thus can neutralize the reactivity of organochlorines in deep aquifers. Other applications of nanoparticles are those that stem from manipulating or arranging matter at the nanoscale to provide better coatings, composites, or additives and those that exploit the particles' quantum effects (e.g., quantum dots for imaging, nanowires for molecular electronics, and technologies for spintronics and molecular magnets).

Nanoparticle Applications in Materials

Many properties unique to nanoparticles are related specifically to the particles' size. It is therefore natural that efforts have been made to capture some of those properties by incorporating nanoparticles into composite materials. An example of how the unique properties of nanoparticles have been put to use in a nanocomposite material is the modern rubber tire, which typically is a composite of a rubber (an elastomer) and an inorganic filler (a reinforcing particle), such as carbon black or silica nanoparticles. For most nanocomposite materials, the process of incorporating nanoparticles is not straightforward. Nanoparticles are notoriously prone to agglomeration, resulting in the formation of large clumps that are difficult to redisperse. In addition, nanoparticles do not always retain their unique size-related properties when they are incorporated into a composite material.

Despite the difficulties with manufacture, the use of nanomaterials grew markedly in the early 21st century, with especially rapid growth in the use of nanocomposites. Nanocomposites were employed in the development and design of new materials, serving, for example, as the building blocks for new dielectric (insulating) and magnetic materials. The following sections describe some of the many applications of nanoparticles and nanocomposites in materials:

- Polymers: Similar to the way in which carbon and silica nanoparticles have been used as fillers in rubber to improve the mechanical properties of tires, such particles and others, including nanoclays, have been incorporated into polymers to improve their strength and impact resistance. In the early 21st century, increasing use of non-petroleum-based polymers that were derived from natural sources drove the development of "all-natural" nanocomposite polymers. Such materials incorporate a biopolymer derived from an alginate (a carbohydrate found in the cell wall of brown algae), cellulose, or starch; the biopolymer is used in conjunction with natural nanoclay or filler derived from the shells of crustaceans. The materials are biodegradable and do not leave behind potentially harmful or non-natural residues.

- Food Packaging: Nanoparticles have been increasingly incorporated into food packaging to control the ambient atmosphere around food, keeping it fresh and safe from microbial contamination. Such composites use nanoflakes of clays and claylike particles, which slow down the ingress of moisture and reduce gas transport across the packaging film. It is also possible to incorporate nanoparticles with apparent antimicrobial effects (e.g., nanocopper or nanosilver) into such packaging. Nanoparticles that exhibit antimicrobial activity had also been incorporated into paints and coatings, making those products particularly useful for surfaces in hospitals and other medical facilities and in areas of food preparation.

- Flame Retardants: Nanoparticles were explored for their potential to replace additives based on flammable organic halogens and phosphorus in plastics and textiles. Studies had suggested that, in the event of a serious fire, products with nanoclays and hydroxide nanoparticles were associated with fewer emissions of harmful fumes than products containing certain other types of additives.

- Batteries and supercapacitors: The ability to engineer nanocomposite materials to have very high internal surface areas for storing electrical charge in the form of small ions or electrons has made them especially valuable for use in batteries and supercapacitors. Indeed, nanocomposite materials have been synthesized for various applications involving electrodes. Composite materials based on carbon nanotubes and layered-type materials, such as graphene, were also researched extensively, making their first appearances in commercial devices in the early 2000s.

- Nanoceramics: A long-term objective in materials science had been to transform ceramics that are brittle and prone to cracking into tougher, more resilient materials. By the early 21st century, researchers had achieved that goal by incorporating an effective blend of nanoparticles into ceramics materials. Other new ceramics materials that were under development included all-ceramic or polymer-ceramic blends, which combined the unique functional (e.g., electrical, magnetic, or mechanical) properties of a nanocomposite material with the properties of ceramics materials.

- Light control: In the 1990s the development of blue light-emitting diodes (LEDs), which had the potential to produce white light at significantly reduced costs, inspired a revolution in lighting. Blue LEDs brought about a need for composite materials that could be used to coat the diodes to convert blue light into other wavelengths (such as red, yellow, or green) in order to achieve white light. One way of obtaining the desired light is by leveraging the size or quantum effect of small semiconducting particles. The application of such particles facilitated the development of nanocomposite polymers for greenhouse enclosures; the polymers optimize plant growth by effectively converting wavelengths of full-spectrum sunlight into the red and blue wavelengths used in photosynthesis. Light conversion in the above cases is achieved with submicron particles of inorganic phosphor materials incorporated into the polymer.

Nanoparticle Applications in Medicine

The small size of nanoparticles is especially advantageous in medicine; nanoparticles can not only circulate widely throughout the body but also enter cells or be designed to bind to specific cells. Those properties have enabled new ways of enhancing images of organs as well as tumours and other diseased tissues in the body. They also have facilitated the development of new methods of delivering therapy, such as by providing local heating (hyperthermia), by blocking vasculature to diseased tissues and tumours, or by carrying payloads of drugs.

Magnetic nanoparticles have been used to replace radioactive technetium for tracking the spread of cancer along lymph nodes. The nanoparticles work by exploiting the change in contrast brought about by tiny particles of superparamagnetic iron oxide in magnetic resonance imaging (MRI). Such particles also can be used to kill tumours via hyperthermia, in which an alternating magnetic field causes them to heat and destroy tissue on a local scale.

Nanoparticles can be designed to enhance fluorescent imaging or to enhance images from positron emission tomography (PET) or ultrasound. Those methods typically require that the nanoparticle be able to recognize a particular cell or disease state. In theory, the same idea of targeting could be used in aiding the precise delivery of a drug to a given disease site. The drug could be carried via a nanocapsule or a liposome, or it could be carried in a porous nanosponge structure and then held by bonds at the targeted site, thereby allowing the slow release of drug. The development of nanoparticles to aid in the delivery of a drug to the brain via inhalation holds considerable promise for the treatment of neurological disorders such as Parkinson disease, Alzheimer disease, and multiple sclerosis.

Nanoparticles and nanofibres play an important part in the design and manufacture of novel scaffold structures for tissue and bone repair. The nanomaterials used in such scaffolds are biocompatible. For example, nanoparticles of calcium hydroxyapatite, a natural component of bone, used in combination with collagen or collagen substitutes could be used in future tissue-repair therapies. Nanoparticles also have been used in the development of health-related products. For example, a sunscreen known as Optisol, invented at the University of Oxford in the 1990s, was designed with the objective of developing a safe sunscreen that was transparent in visible light but retained ultraviolet-blocking action on the skin. The ingredients traditionally used in sunscreens were based on large particles of either zinc oxide or titanium dioxide or contained an organic sunlight-absorbing compound. However, those materials were not satisfactory: zinc oxide and titanium dioxide are very potent photocatalysts, and in the presence of water and sunlight they generate free radicals, which have the potential to damage skin cells and DNA (deoxyribonucleic acid).

Scientists proceeded to develop a nanoparticle form of titanium oxide that contained a small amount of manganese. Studies indicated that the nanoparticle-based sunscreen was safer than sunscreen products manufactured by using traditional materials. The improvement in safety was attributed to the introduction of manganese, which changed the semiconducting properties of the compound from n-type to p-type, thus shifting its Fermi level, or oxidation-reduction properties, and making the generation of free radicals less likely.

Treatments and diagnostic approaches based on the use of nanoparticles are expected to have important benefits for medicine in the future, but the use of nanoparticles also presents significant challenges, particularly regarding impacts on human health. For example, little is known about the fate of nanoparticles that are introduced into the body or whether they have undesirable effects on the body. Extensive clinical trials are needed in order to fully address concerns about the safety and effectiveness of nanoparticles used in medicine. There also are manufacturing problems to be overcome, such as the ability to produce nanoparticles under sterile conditions, which is required for medical applications.

Manufacture of Nanoparticles

Nanoparticles are made by one of three routes: by comminution (the pulverization of materials), such as through industrial milling or natural weathering; by pyrolysis (incineration); or by sol-gel synthesis (the generation of inorganic materials from a colloidal suspension). Comminution is known as a top-down approach, whereas the sol-gel process is a bottom-up approach. Examples of those three processes (comminution, pyrolysis, and sol-gel synthesis) include the production of titania nanoparticles for sunscreens from the minerals anatase and rutile, the production of fullerenes or fumed silica (not to be confused with silica fume, which is a different product), and the production of synthetic (or Stöber) silica, of other "engineered" oxide nanoparticles, and of quantum dots. For the generation of small nanoparticles, comminution is a very inefficient process.

Detection, Characterization and Isolation

The detection and characterization of nanoparticles present scientists with particular challenges. Being of a size that is at least four to seven times smaller than the wavelength of light means that individual nanoparticles cannot be detected by the human eye, and they are observable under optical microscopes only in liquid samples under certain conditions. Thus, in general, specialized techniques are required to see them, and none of those approaches is currently field-deployable.

Techniques to detect and characterize nanoparticles fall into two categories: direct, or "real space," and indirect, or "reciprocal space." Direct techniques include transmission electron microscopy (TEM), scanning electron microscopy (SEM), and atomic force microscopy (AFM). Those techniques can image nanoparticles, directly measure sizes, and infer shape information, but they are limited to studying only a few particles at a time. There are also significant issues surrounding sample preparation for electron microscopy. In general, however, those techniques can be quite effective for obtaining basic information about a nanoparticle.

Indirect techniques use X-rays or neutron beams and obtain their information by mathematically analyzing the radiation scattered or diffracted by the nanoparticles. The techniques of greatest relevance to nanoscience are small-angle X-ray scattering (SAXS) and small-angle neutron scattering (SANS), along with their surface-specific analogues GISAXS and GISANS, where GI is "grazing incidence," and X-ray or neutron reflectometry (XR/NR).

The advantage of those techniques is that they are able to simultaneously sample and average very large numbers of nanoparticles and often do not require any particular sample preparation. Indirect techniques have many applications. For example, in studies of nanoparticles in raw sewage, scientists used SANS measurements, in which neutrons readily penetrated the turbid sewage and scattered strongly from the nanoparticles, to follow the aggregation behaviour of the particles over time. The isolation of nanoparticles from colloidal and larger matter involves specialized techniques, such as ultra centrifugation and field-flow fractionation. Such laboratory-based techniques are normally coupled to standard spectroscopic instrumentation to enable particular types of chemical characterization.

Nanoparticles in the Environment

Nanoparticles occur naturally in the environment in large volumes. For example, the sea emits an aerosol of salt that ends up floating around in the atmosphere in a range of sizes, from a few nanometres upward, and smoke from volcanoes and fires contains a huge variety of nanoparticles, many of which could be classified as dangerous to human health. Dust from deserts, fields, and so on also has a range of sizes and types of particles, and even trees emit nanoparticles of hydrocarbon compounds such as terpenes (which produce the familiar blue haze seen in forests, from which the Great Smoky Mountains in the United States get their name).

Human-made (anthropogenic) nanoparticles are emitted by large industrial processes, and in modern life it is particles from power stations and from jet aircraft and other vehicles (namely, those powered by internal-combustion engines; car tires are also a factor) that constitute the major fraction of nanoparticle emissions. Types of nanoparticles that are emitted include partially burned hydrocarbons (in soot), ceria (cerium oxide; from vehicle exhaust catalysts), metallic dust (from brake linings), calcium carbonate (in engine lubricating oils), and silica (from car tires). Other sources of nanoparticles to the environment include the semiconductor industry, domestic and industrial wastewater discharges, the health care industry, and the photographic industry. However, all those emission levels are still considered to be lower than the levels of nanoparticles produced through natural processes. Indeed, recent human-made particles contribute only a small amount to air and water pollution.

Understanding the relationship between nanoparticles and the environment forms an important area of research. There are several mechanisms by which nanoparticles are believed to affect the environment negatively. Two scenarios that are under investigation are the possibilities that:

- The mobility and sorptive capacity of nanoparticles (natural or human-made) make them potent vectors (carriers) in the transport of chemical pollutants (e.g., phosphorus from sewage and agriculture), particularly in rivers and lakes.
- Some nanoparticles are able to reduce the functioning of (and may even disrupt or kill) naturally occurring microbial communities, as well as microbial communities that are employed in industrial processes (e.g., those that are used in sanitation processes, including sewage treatment).

Nanoparticles also can have beneficial impacts on the environment and appear to contribute to natural processes. Thus, in addition to the potential use of nanoparticles to remove chemical contaminants from the environment, scientists are investigating how nanoparticles interact with all life-forms from fungi to microbes, algae, plants, and higher-order animals. That type of study is essential not only to improving scientists' knowledge of nanoparticles but also to gaining a more complete understanding of life on Earth, since the soil is naturally full of nanoparticles, in a richly diverse environment.

Health Effects of Nanoparticles

Humans have evolved to cope with most naturally occurring nanoparticles. However, some nanoparticles, generated as a result of certain human activities such as tobacco smoking and fires, account for many premature deaths as a result of lung damage. For example, fires from the types of cooking stoves used in developing countries are known to emit fine particles and lead to early mortality, especially among women who routinely work near the stoves.

Laboratory and clinical investigations of the effects of nanoparticles on health have been somewhat controversial and remain largely inconclusive. Most studies in animals have involved nanoparticle inhalation, and the dosages have been very large. The results of those studies have indicated that large quantities of nanoparticles can cause cellular damage in the lungs, with lung cells absorbing the particles and becoming damaged or undergoing genetic mutation. Animal studies involving the ingestion of nanoparticles in food or water suggest that nanoparticles can also affect health in other ways. For example, consumption of the food additive E171, which consists of titanium dioxide nanoparticles, is associated with changes in gut microbiota (bacteria occurring in the gut), potentially contributing to the development of conditions such as inflammatory bowel disease.

In humans, the health effects of typical exposure levels those that are encountered by most persons during daily activities remain unknown. Nonetheless, there is a general awareness of the problems that might occur upon excess exposure to nanoparticles, and, thus, most manufacturers of such particles take serious precautions to avoid exposure of their workers. Efforts have been made to educate the public in the use of nanoparticle-containing products. The existence of pressure groups has also helped to ensure nanoparticle safety compliance among manufacturers. However, nanoparticles offer tremendous potential for new or improved forms of health care treatment. That has spawned a new field of science called nanomedicine.

Nanoporous Materials

Classically, porous matter is seen as a material that has voids through and through. The voids show a translational repetition in 3-D space, while no regularity is necessary for a material to be termed "porous." A typical and relatively simple porous system is one type of dispersion classically described in colloid science, namely a foam or, better, a solid foam. In correlation with this, the most common way to think about a porous material is as a material with gas-solid interfaces as the most dominant characteristic. This already indicates that classical colloid and interface science as the creation of interfaces due to nucleation phenomena (in this case nucleation of wholes), decreasing interface energy, and stabilization of interfaces is of elemental importance in the formation process of nanoporous materials.

These factors are often omitted because the final products are stable. Indeed, they are just metastable. This metastability is due to the rigid character of the void-surrounding network, which is covalently cross-linked in most cases. However, it should be noticed that most of the porous materials reported in the literature or those that are of high technical relevance are not stable by thermodynamic means. As soon as kinetic energy boundaries are overcome, materials start to break down. One example should elucidate this. Porous silica, for instance, is just metastable. As soon as the temperature is raised and the melting point is reached, primary particles in the network begin to fuse and it comes to phase separation into a nonporous silica phase. Finally, at very high temperatures, the thermodynamic stable phase of SiO_2 quartz emerges.

Control over interface energy and metastabilization of nanodimensional holes becomes of special importance when the task is to produce nanoporous materials. Here, the whole range of pores from 1 nm to 1000 nm is interpreted as nanoporous materials. Additionally, we will skip the strict limitation of just regarding 3-D distributed solid-gas interfaces as porous materials. Instead, when the pores happen to be filled with something the materials will also be judged to be porous as long as it is possible to remove this filling. The pores might be filled by any matter that is different from the surrounding network. However, it should be noted that nanoporous materials with filled pores might be more correctly described as nanostructured materials. Accordingly, porous materials might be classified by the size of pores or may be distinguished by different network materials.

Relevance of Nanoporous Materials

One can find a whole variety of nanoporous materials in nature executing many different functions. The most common task for nanoporous materials in nature is to make inorganic material much lighter while preserving or improving the high structural stability of these compounds. Often, by filling the voids between inorganic matter the desired properties of the hybrid materials exceed the performances of the pure compounds by several orders of magnitude. Nanoporous materials in nature are organic inorganic hybrids. Naturally occurring materials exhibit synergistic properties. Neither the organic material filling the void nor the inorganic network materials are able to achieve comparable performances by themselves. Examples of these materials are found for diatoms, radiolaria, or the famous abalone shell. Even in the human body, life without bones would not be conceivable.

It is seen that complex mechanisms are involved in the formation of these hierarchical materials. Similar to the structure motives on different length-scales cells, vesicles, supramolecular structures, and biomolecules are involved in the structuring process of inorganic matter occurring in nature. This process is commonly known as biomineralization. It is often not seen in this relation, but it will be shown later that ordered porous materials, and therefore artificial materials, are constructed according to very similar principles.

A completely different area where nanoporous materials are highly important is in the lungs, where a foam with a high surface area permits sufficient transfer of oxygen to the blood. Even the most recent developments in nanoporous materials, such as their application as photonic materials, are already present in nature; the color of butterfly wings, for instance, originates from photonic effects.

It can be concluded that nature applies the concept of nanoporous materials (either filled or unfilled) as a powerful tool for constructing all kinds of materials with advanced properties. It is therefore not surprising how much research has recently been devoted to porous materials in different areas such as chemistry, physics, and engineering. The current interests in nanoporous materials are now far behind their size-sieving properties.

Classifications

Classification by Pore Size

The pore-size regime for nanoporous materials ranges from the 1-nm region to 1000 nm (=1 μm). According to IUPAC, three distinctions can be made:

- Microporous materials 0–2-nm pores.
- Mesoporous materials 2–50-nm pores.
- Macroporous materials >50-nm pores.

A comparison of these different pore systems is given in figure. However, in these cases ordered pore systems are shown. Order is no premise at all. Microporous, mesoporous, or macroporous materials might have random systems of pores. In the field of mesoporous materials in particular, the mistake is often made of mixing up mesoporous and ordered materials.

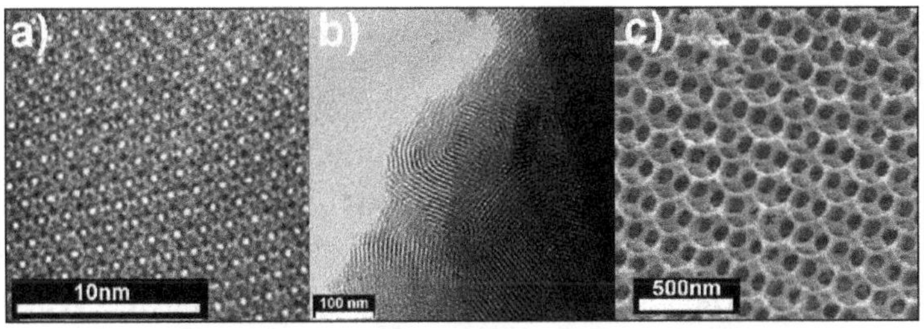

Three different types of nanoporous silica materials. (a) A microporous material: zeolite Y. (b) A mesoporous materials. (c) A macroporous material.

Ordered porous materials are judged to be much more interesting because of the control over pore sizes and pore shapes. Their disordered counterparts exhibit high polydispersity in pore sizes, and the shapes of the pores are irregular. Ordered porous materials seem to be much more homogeneous. But one has to be careful with this statement because it is often led by aesthetic arguments caused by the beauty of images. In many cases a material possesses more than one porosity. This could be:

- For microporous materials: An additional meso- or macroporosity caused by random grain packing.
- For mesoporous materials: An additional macroporosity caused by random grain packing, or an additional microporosity in the continuous network.
- For macroporous materials: An additional meso- and microporosity.

These factors should be taken into consideration when materials are classified according to their homogeneity. A material possessing just one type of pore, even when the pores are disordered, might be more homogenousthan one having just a fraction of nicely ordered pores. A good tool for deciding and classifying porous materials according to the size of their pores is sorption analysis. Indeed, pores of different sizes lead to totally different characteristics in sorption isotherms.

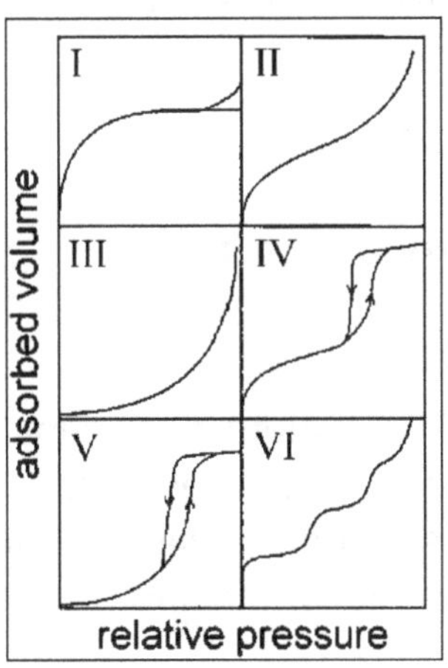

Sorption isotherms of microporous materials (I), mesoporous materials (V), and macroporous materials (III).

The correlation between the vapor pressure and the pore size is given by the Kelvin equation (shown here in a slightly modified form):

$$r_p\left(\frac{p}{p_o}\right) = \frac{2\gamma V_L}{RT\ln(p/p_o)} + t\left(\frac{p}{p_o}\right)$$

where $r_p \cong$ pore radius, $\gamma \cong$ surface tension, $t \cong$ thickness of the adsorbate film, $V_L \cong$ molecular volume of the condensate.

Therefore, the isotherms of microporous materials show a steep increase at very low pressures (relative pressures near zero) and reach a plateau quickly. Mesoporous materials are characterized by a so-called capillary condensation step and a hysteresis (a difference between adsorption and desorption). Macroporous materials show a single or multiple adsorption steps near the pressure of the standard bulk condensated state (relative pressure approaches one).

Classification by Network Material

One of the most important goals in the field of nanoporous materials is to achieve any possible chemical composition in the network materials "hosting" the pores. It makes sense to divide the materials into two categories: inorganic materials and organic materials. Among the inorganic material, which is the larger group, we find:

- Inorganic oxide-type materials. This is the field of the most commonly known porous silica, porous titania, and porous zirconia materials.
- A category of its own is given for nanoporous carbon materials. In this category are the highly important active carbons and some examples for ordered mesoporous carbon materials.
- Other binary compounds such as sulfides, nitrides etc. Into this category also fall the famous $AlPO_4$ materials.
- There are already some examples in addition to carbon where just one element (for instance, a metal) could be prepared in a nanoporous state. The most prominent member of this class of materials is likely to be nanoporous silicon, with its luminescent properties.

There are far fewer examples of nanoporous organic materials, such as polymers.

Modern Analysis of Nanoporous Materials

Gas Adsorption Methods

Gas sorption represents a widely used technique for characterizing micro- and mesoporous materials and provides porosity parameters such as pore size distributions, surface areas, and pore volumes. In the following, a brief description of classical methods of analyzing sorption data is followed by an overview of recent advances in the interpretation of sorption experiments.

In a typical sorption experiment the uptake of gases such as nitrogen, krypton, and CO_2 is measured as a function of relative pressures $p/p_0 < 1$ at constant temperature. p and p_0 are the equilibrium vapor pressures of the liquid in the pores and that of the bulk liquid, respectively. The interaction between the pore walls and the adsorbate is based on physisorption (van der Waals interaction) and leads to the formation of adsorbate layers at low p/p_0. Simplistically, the macroscopic laws of classical thermodynamics predict that the confinement of pores with radii on the nanometer scale leads to the condensation of gas inside the pores at a pressure smaller than p_0. In a typical sorption experiment the adsorbed volume is plotted versus p/p_0, and this "sorption isotherm" is the superposition of different uptake mechanisms. At low p/p_0 adsorption in micropores takes place, which is supposed to be a process of volume filling rather than capillary condensation. The shape of an isotherm itself distinguishes between representative types of nanoporous materials, based on the classifications by IUPAC or de Boer.

The isotherms of microporous materials are characterized by a steep increase of the isotherm at low p/p_0, ending up in a plateau at larger p/p_0. In a mesoporous substrate, with increasing values of p/p_0, a liquid-like adsorbate lm of statistical thickness $t\ (p/p_0)$ is formed on the pore walls. At a certain pressure, capillary condensation takes place, filling the mesopores with liquid, which is apparent in isotherms as a pronounced increase in the adsorbed amount. The total pore volume ("porosity") is given by the overall uptake of adsorbate. Zsigmondy proposed the first explanation of the capillary condensation in a single in finite cylindrical mesopore based on the macroscopic Kelvin equation $\ln(p/p_0) = 2\sigma V_L / RTr_m$, where V_L and σ are the molal liquid volume and the liquid-gas surface tension at temperature T, respectively, and r_m is the mean radius of curvature of the liquid/gas interface (cylindrical for the condensation, spherical for desorption).

Based on the classical treatment of Cohan and the Kelvin equation, the condensation of a liquid in a nanoporous material at a certain p/p_0 can be related to the corresponding mesopore size, thus also providing a pore size distribution (PSD). Since the condensation starts at a relative pressure p/p_0, where the walls are covered by a film of thickness $t\ (p/p_0)$, in the so-called modified Kelvin equation $2/r_m$ is replaced by $f/(r - t(p/p_0))$, where r is the "true" mesopore radius and *f* is the meniscus shape factor, which is 1 or 2 for the filling or emptying of the mesopore, respectively. This procedure represents the basis of the well-known "Barrett-Joyner-Halenda (BJH)" method, currently the procedure most frequently used to determine PSDs. The dependence $t\ (p/p_0)$ can be described by the approach of Frenkel-Halsey-Hill describing the sorption on nonporous silica. An improved treatment, originating with Derjaguin, takes into account the influence of surface forces on adsorbed film equilibrium and stability, which leads to predictions for capillary condensation and desorption pressures that are substantially different from those of Cohan's theory. In addition, the pore geometry significantly affects thermodynamic properties of confined fluids and their adsorption behavior. The confinement is stronger in spherical pores compared with cylindrical pores of the same diameter, leading to a shift of the capillary condensation to lower p/p_0; therefore the aformentioned procedures are restricted to cylindrical pores. Broekhoff and de Boer described the condensation in spherical ink-bottle pores, connected by narrow cylindrical windows, based on the Kelvin-Cohan approach, also predicting the appearance of hysteresis.

The progress in synthesizing mesoporous materials with well-defined pore morphologies in terms of uniform mesopore sizes and pore shapes allowed the testing, optimization, and further development these classical approaches. Based on the BJH method, Kruk and Jaroniec determined the mesopore size of MCM-41 by using the "KJS" approach, which corrects the Kelvin equation by an empirical additive constant (0.3) for the mesopore radius, which, however, does not have a theoretical foundation: $r(p/p_0)[nm] = 2\sigma V_L / RT\ \ln(p_0/p) + t(p/p_0) + 0.3$. Moreover, Kruk and Jaroniec used basic geometrical relationships to determine the mesopore diameter of MCM-41 based on the (100) interplanar spacing as determined from X-ray diffraction and the mesopore volume, assuming an array of uniform pores.

A further approach to calculating pore volumes and mesopores sizes is based on the Gurvich approach, in which the adsorbed volume $V_{0.4}$ at $p/p_o \approx 0.4$ is used to calculate the pore diameter $D = 4V_{0.4} / S_{BET}$, where S_{BET} is the specific surface area obtained from the method introduced by Brunauer, Emmet, and Teller (BET). The BET method is based on the assumption that multiplayer formation takes place prior to capillary condensation and that the equilibrium state is characterized by different rate constants for adsorption and desorption for the mono- and multilayer. The adsorbed amount n_a, p/p_o, and the monolayer capacity n_m are related by the BET equation, from which the specific surface area S_{BET} is calculated as $S_{BET} = n_m a_m N_A$, where a_m is the area of an adsorbate molecule and N_A is Avogadro's number.

Further widely used methods for determining structural parameters are comparative plots such as the t-plot and α-plot methods. Typically, the amount adsorbed on the porous solid under study is plotted as a function of the amount adsorbed on an ideally nonporous reference solid with similar surface characteristics, providing parameters such as the overall pore volume, specific surface area, and micropore volumes. This procedure has been used to determine microporosity in mesoporous silicas such as SBA-15 showing nonnegative intercepts in the comparative plot. Based on comparative plots, researcher concluded that MCM-41 does not contain extra microporosity.

The procedures described above are not appropriate for the characterization of microporous materials, particularly microporous carbons and zeolites. Among the most frequently used evaluation procedures are the phenomenological models based on Dubinin's theory of volume filling of micropores, such as the Dubinin-Radushkevich (DR), Dubinin-Astakhov and DubininStoeckli equations. Another approach is the Saito-Foley method, which is an extension of the Horvath-Kawazoe method.

It has turned out that the macroscopic thermodynamics of the classical methods described above do not provide reliable descriptions of materials with mesopore sizes below about 4 nm for oxidic materials. One of the main shortcomings of these approaches lies in the non-consideration of fluid-wall interactions. Recent progress in understanding capillary condensation deals with molecular level models. The methods of the grand canonical Monte Carlo (GCMC) simulations, molecular dynamics, and density functional theory (DFT) allow direct modeling of capillary condensation/desorption phase transitions and are capable of generating hysteresis loops of simple fluids sorbed in model pores.

Neimark and Ravikovitch have shown that the nonlocal density functional theory (NLDFT), with properly chosen parameters of fluid-fluid and fluid-solid intermolecular interactions, quantitatively predicts desorption branches of hysteretic isotherms of nitrogen and argon on reference MCM-41 samples with pore channels narrower than 5 nm. This method was tested against Monte Carlo simulations and was shown to provide reliable pore sizes and wall thicknesses in MCM-41 materials. Neimark and Ravikovitch extended the NLDFT approach to MCM-41- type materials such as SBA-15 with pore sizes larger than 5 nm, and, in a recent study, they applied the NLDFT approach successfully to cage-like mesopore silicas such as SBA-1, SBA-2, and SBA-12. Moreover, the NLDFT predictions of equilibrium and spontaneous capillary condensation transitions for pores wider than 6 nm were well approximated by the macroscopic equations of the Derjaguin-Broekhoff-de Boer theory, while the results of the traditional Cohan equation (BJH method) were shown to be significantly in error. In spite of the recent progress in the theoretical understanding of sorption phenomena, certain issues of sorption are still unclear:

- The nature of the hysteresis in mesoporous materials is still subject to intensive theoretical and experimental research and is not yet fully understood. While certain materials with small mesopores such as MCM-41 or MCM-48 exhibit equilibrium capillary condensation, not showing a hysteresis loop at all in nitrogen sorption experiments at T = 77 K, mesoporous silicas like SBA-15 show pronounced hysteresis loops with parallel adsorption and desorption branches. In contrast, other types of mesoporous silicas show a steep decrease in the desorption curve. Recent NLDFT studies were able to shed some more light on the hysteresis phenomenon. It was concluded that in the range of pore sizes greater than 5 nm, for MCM-41-like materials the experimental desorption branch corresponds to equilibrium evaporation, while the capillary condensation branch corresponds to spontaneous (spinodal) condensation. Similar results were obtained for cage-like pores in siliceous materials. Froba and co-workers carried out a systematic study of adsorption/desorption phenomena in MCM-48 by using different gases such as nitrogen, argon, and krypton at various temperatures. It was concluded that hysteresis is a function of the pore size and the temperature.
- The simultaneous presence of mesopores and irregular micropores in the walls, as shown for SBA-15 and other silicas obtained from PEO-containing templates, severely impedes the evaluation of surface areas and the quantification of the micropore volumes. Although it was shown that the use of comparative plots underestimates the microporosity, t-plots and α-plots are still frequently used for these materials.
- The dependence t (p/p_0) is still a matter of discussion because of the lack of independent techniques for determining film thicknesses, and usually reference data from nonporous materials are used. An exact knowledge of t (p/p_0) is needed for the determination of mesopore sizes by classical methods and for testing DFT models. Researcher have determined $t(p/p_0)$ by a combination of nitrogen sorption and small-angle neutron scattering.

In mercury porosimetry (MP), gas is evacuated from the sample, which is then immersed in mercury, and an external pressure is applied to gradually force the non-wetting mercury into the sample. By monitoring the incremental volume of mercury intruded for each applied pressure, the pore size distribution of the sample can be estimated in terms of the volume of the pores intruded for a given diameter D. The evaluation of pore sizes from MP is based on the Washburn equation, quantifying the pressure p required to force a nonwetting fluid into a circular cross-sectional capillary of diameter D. MP allows the determination of PSD between 3 nm and 200 nm and is therefore inappropriate for microporous materials but more suitable for pore sizes above ca. 30 nm compared with nitrogen sorption. As further parameters, the total pore volume V_{tot} is accessible from the total intruded volume of mercury at the highest pressure determined, and the total pore surface S is calculated from $S = (1/\gamma|\cos\theta|)\int_0^{V_{tot}} p\,dV$. MP has inherent shortcuts in determining mesopore sizes, especially below 20 nm. During the measurement, high pressures used to force mercury into small pores may compress the sample. Damage or compression of highly porous silica has been reported previously. In addition, mercury porosimetry overestimates the volume of the smallest pores in the case of ink-bottle-shaped pores by the small openings.

Electron Microscopy

Electron microscopy (transmission electron microscopy (TEM) for micro- and mesoporous samples and scanning electron microscopy (SEM) for macroporous samples) is an indispensable tool for the investigation of porous materials. The biggest advantage of these techniques is that they deliver an optical image of the samples.

Diffraction Techniques

Experiments using elastic X-ray and neutron scattering have turned out to be an invaluable tool for the characterization of various types of porous materials, providing quantitative parameters such as the pore size, surface area, and pore volume. In addition, diffraction techniques allow the determination of the shape and, in particular, the spatial distribution of the pores, for both highly ordered arrays such as in MCM-41 and a more disordered arrangement of pores, for instance in activated carbons. In the early 1900s, Max von Laue, W. L. Bragg, W. H. Bragg, and others laid the groundwork for X-ray crystallography, which has become a powerful method of visualizing complex inorganic and organic crystalline materials. In spite of the variety of different diffraction techniques currently available, they all are based on the same physical phenomenon, namely the scattering of X-rays and neutrons by the atoms through their electrons or nuclei, respectively.

In the Fraunhofer approximation, the interaction of X-rays with electrons leads to the superposition of the coherent scattering of the basic scattering centers (atoms, molecules, or pores), without changing the energy of the incoming X-rays ("elastic scattering"). The resulting coherent scattering pattern is directly related to the mutual position, size, and scattering power of these scattering units. For almost all types of nanoporous materials, no single crystal diffraction datum is obtainable, not even for the crystalline zeolites. Therefore, most of the most prominent types of nanoporous materials (zeolites, highly ordered mesoporous materials such as MCM-41 or SBA-type materials, and porous carbons) are studied in diffraction experiments as polycrystalline powders. In this case, the powder diffraction raw data are obtained as 1D plots of the coherent scattering intensity versus the scattering angle 2θ. Only in the case of thin porous films with an oriented alignment of the pores relative to the substrate 2D diffraction can patterns be obtained.

The main problem in analyzing scattering patterns of any kind of polycrystalline nanoporous materials lies in the extraction of a maximum of structural information: the highly advanced crystallographic strategies for the interpretation of 3D diffraction patterns of single crystals are not applicable, requiring different evaluation approaches. Depending on the length scale of the pore size, two methods can be distinguished regarding both the experimental realization and the theoretical treatment. Basically, the characteristic length scale d describing the pore system and the corresponding diffraction angle 2θ are related by the Bragg equation, $s = 1/d = (2\sin\theta)/\lambda$, where λ is the wavelength and s is the corresponding scattering vector. Several excellent textbooks are available on the general principles of diffraction theory and experiments.

Zeolites and other crystalline porous materials with pore sizes below the nanometer scale, which are obtained as polycrystalline powders, are investigated by wide-angle scattering (WAS) techniques, for instance by standard powder diffractometers, which are usually equipped with a 1D detector. The WAS scattering from a polycrystalline porous material is given by $I(|\vec{s}|) = \left\langle |Z(\vec{s})|^2 |F(\vec{s})|^2 \right\rangle_\omega$

where $Z(\vec{s})$ represents the 3D spatial distribution of the unit cell ("lattice factor"), and the electron distribution within the unit cell is described by the "atomic form factor" $F(\vec{s})$. The main difficulty in the interpretation of diffraction patterns of polycrystalline material arises from the spatial average $\langle ... \rangle_\omega$, which leads to a loss of information compared with single crystals. Once the crystal structure has been solved, the pore size and shape are obtained from the atom coordinates.

Substantial progress has been achieved in the structure determination of zeolites based on powder diffraction data, which may comprise up to 200 reflection peaks in the wide-angle diffraction region. For instance, in the FOCUS software developed by Grosse-Kunstleve and co-workers, an automatic Fourier recycling algorithm is combined with a specialized topology (framework) search specific to zeolites. In essence, this algorithm takes into account chemical information on the composition and structure that is common to all zeolite frameworks. For instance, the evaluation algorithm includes the feature that these open-framework structures can be described as 3D four-connected nets of tetrahedrally coordinated atoms. Recently this concept was applied successfully to various types of zeolites, such as ZSM-5 and cloverite.

Compared with zeolites, nanoporous materials with pores sizes above 1 nm are characterized by a more disordered spatial distribution of the pores and are studied by so-called small-angle scattering (SAS), with the use of either X-rays (SAXS) or neutrons (SANS). The evaluation of SAS data for porous materials is usually based on the approximation that the material can be regarded as a so-called two-phase system: in the case of X-ray scattering, basic theoretical considerations show that the SAS of such materials arises from the scattering at the void-solid interface and is related to the scattering contrast $(\delta_1 - \delta_2)^2$, where δ_1 is the average electron density of the voids ($\delta_1 = 0$) and δ_2 represents the average electron density of the solid. The SAS of a sufficiently ordered nanoporous material with pores of a distinct shape is also given by $I(|\vec{s}|) = \left\langle |Z(\vec{s})|^2 |F(\vec{s})|^2 \right\rangle_\omega$, where the lattice factor Z has the same meaning as in wide-angle scattering, and the form factor F here corresponds to the shape of the mesopore/solid rather than the electron density of single atoms.

Polycrystalline mesoporous materials with a distinct pore shape and a well-defined 3D or 2D alignment, realized in silicas of the MCM or SBA families, give rise to SAS patterns with a characteristic sequence of reflections ("peaks"). The pore structures of even the most highly ordered mesoporous materials show a considerably lower order and symmetry compared with zeolites, leading to a much smaller number of possible mesostructures and corresponding SAS patterns. Therefore, the SAS patterns of this limited number of possible regular mesopore structures and space groups can serve as "fingerprints," allowing an almost unambiguous assignment of a certain mesopore lattice structure, if a sufficient number of reflection peaks are obtained. In combination with TEM, SAXS experiments turned out to be a powerful technique for determining the alignment and structure in a variety of mesoporous MCM- and SBA- type materials and other structures.

As the main information, SAXS provides the crystallographic space group of the mesopore arrangement and the corresponding lattice parameter. While the 3D alignment of the mesopores can be obtained from the bare SAS peak positions, the determination of mesopore sizes from single peak analyses involves substantial uncertainties, because both the peak proles and intensities can be substantially superimposed by various factors such as smearing, incoherent background scattering, the arrangement of the mesopores and background fluctuations, and the presence of additional intrawall micropores. In particular, so far no satisfactory approach has been developed to quantitatively simulate the influence of 2D and 3D disorder on SAS.

Even highly ordered mesopore systems such as MCM-41 show a certain 2D displacement of the cylindrical mesopores on the hexagonal lattice, which together with the polydispersity of the pores may result in a non-negligible overlap of the peaks, thus severely aggravating a meaningful pore size analysis. In recent studies, the SAS of MCM-41 and SBA-15 silicas was modeled and compared with experimental data; however, the peak analyses did not take into account the impacts mentioned above. Figure shows typical SAXS patterns of three representative silicas exhibiting different degrees of structural order and regularity. Curve D in figure is the SAXS curve of SBA-2 silica, which has a hexagonal, close-packed structure.

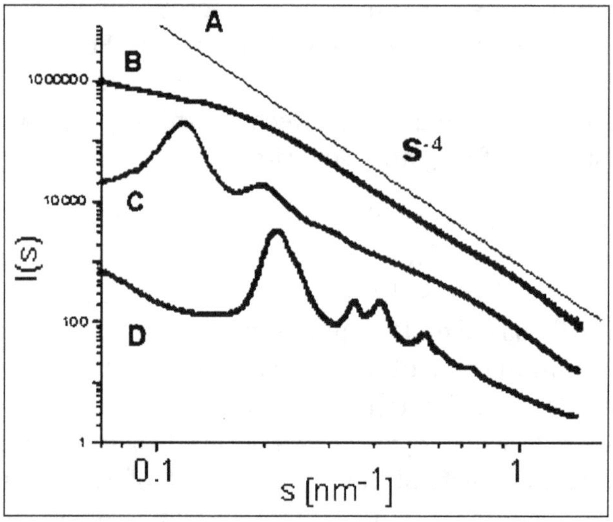

SAXS curves of various porous silicas. (A) Theoretical Porod asymptote for two-phase systems. (B) Poroussilica obtained from a PEO template. (C) Mesoporous SE1010 silica. (D) SBA-2.

Aside from the mesoporous materials with a high degree of order in terms of a regular spatial distribution and a high uniformity of pore size and shape, SAS is also applied to mesopores showing a certain disorder in the mesopore alignment and a pronounced polydispersity of pore sizes, which leads to SAS patterns with only few broad reflections. Researcher showed that the SAS of this type of material, possessing only a "liquid like" local ordering of a system of polydisperse mesopores, can be reasonably approximated with the use of hard-sphere structure factors such as the Percus-Yevick approach. Curve C in figure represents the SAXS curve of a mesoporous silica obtained from PS-b-PEO templates ("SE1010 silica"). It is noteworthy that the SAXS data of both SBA-2 and SE1010 silica obey the Porod law, $\lim_{s\to\infty} s^4 I(s) = \frac{k}{2\pi^3 l_p}$, which is theoretically predicted for a two-phase system, where l_p is the so-called Porod length and k is the Porod invariant. Curve B in figure shows the SAXS curve of a porous silica with a disordered pore structure, templated with poly(ethylene oxide), also exhibiting the Porod asymptote at large values of the scattering vector s. It is obvious from figure, especially from curves B and C, that different approaches are needed for the evaluation of SAXS data from pore systems showing a lower degree of uniformity.

A further challenge in the interpretation of SAS data from mesoporous materials is the presence of additional micropores (for example in SBA-15), located in the mesopore walls and created by the PEO chains after template removal. Researcher. analyzed the SAXS data of SBA-15, assuming a "corona" of lower silica density around the mesopores caused by the intrawall micropores. However, this approach suffers from the general problems of single-peak analyses.

A different approach for the evaluation of SAS data, without assuming a specific model, was recently pursued by using the concept of the so-called chord-length distribution (CLD) g(r). g(r) is a statistical function describing the probability of finding a chord of length r that is a connector of two points on the solid-void interface. The only pre-assumption of this concept is the formal description of the pore system as a two-phase system, but no assumptions about the pore shape and distribution are needed. Therefore the CLD approach is most appropriate for disordered pore systems. The CLD evaluation method provides various useful structural parameters, such as the Porod length l_p (which is the first moment of g(r)), from which an average pore size l_{pore} and wall thickness l_{wall} can be calculated by $l_{pore} = l_p / (1-\phi)$ and $l_{pore} = l_p / (1-\phi)$, where ϕ is the volume fraction of voids. Furthermore, the inner surface area per unit volume S/V is related to l_p via $l_p = 4\phi(1-\phi)V = S$. By applying the CLD method to mesoporous silicas, it was shown that PEO-containing templates result in a substantial degree of microporosity. In addition, this approach allowed the determination of the distribution of the size of the micropore, which was found to be ca. 0.8–1.8 nm in diameter.

A new and interesting experimental SAS technique for the investigation of mesoporous silica was introduced and developed by Hoinkis at the Hahn-Meitner Institute, Berlin. In this experimental setup, SANS is combined with in situ nitrogen sorption at T = 77 K. In essence, complete SANS curves are obtained during nitrogen sorption at arbitrary points on the isotherm, thus allowing the investigation of the subsequent steps of pore filling as a function of the pore size. The experiment takes advantage of the fortunate situation that the scattering contrast between amorphous silica and condensed nitrogen (at T = 77 K) is almost zero. Based on this method, the pore structure of various types of mesoporous silicas was studied. In a similar experiment, Albouy combined a nitrogen sorption experiment with in situ SAXS, which faces certain problems in the data interpretation because of the situation of a multiphase system with respect to the electron densities of silica, liquid nitrogen, and voids.

SAS techniques are also applied intensely to elucidate microporous carbons. In this case, the pore structure is characterized by a significant degree of polydispersity in pore size and shape and an irregular spatial distribution of the micropores. Various evaluation methods have been used to determine the micropore size from SAXS. Whereas in certain studies the evaluation is based on a dilute system of polydisperse spheres, the application of the CLD concept has shown that the micropores in carbons are more consistent with an accute, needle-like shape. In addition, recent detailed SAXS studies indicate that the apparent fractal structures reported for certain microporous carbons are due to a misinterpretation of SAS data. Instead, Ruland and others have shown that the CLD concept provides an appropriate method for elucidating the microporous structure of carbons both qualitatively and quantitatively. Based on this method, the changes in the porosity that occur with typical physicochemical processing such as thermal treatment could be described by various structural parameters such as the average pore size, wall thickness, and angularity of the micropores. For example, SAS experiments revealed that the thermal treatment of carbons above ca. 1200 °C leads to an increase in the average pore size. In a recent study, the changes in the void-solid microstructure due to these treatments were studied by a combination of SAXS and WAXS.

Positron Annihilation

In the past 20 years, positron annihilation lifetime spectroscopy (PALS) has been developed into a powerful tool for the detection and quantification of defects on the atomic scale in various types of solids. PALS is sensitive to different kinds of defects, such as dislocations and vacancies in metals or crystals, grain boundaries, and voids and pores. Similar to scattering techniques, PALS is a non-invasive technique and thereby allows the detection of inaccessible pores. In the area of micro- and mesoporous materials, PALS is predominantly applied to porous polymers and thin porous films. PALS is based on the decay of positrons into two γ photons ("annihilation") and has been described in various publications.

With ^{22}Na as the radioactive source, the formation of positrons (β^+) by radioactive decay is accompanied by the simultaneous emergence of a γ-quantum of 1.273 MeV, which defines the starting signal of the positron lifetime measurement. Entering the sample, the positrons lose their high energy by inelastic collisions with electrons. These "thermalized" positrons have energies on the order of a few meV, form positroniums (Ps, the electron-positron bound state), and diffuse through the solid until annihilation after their specific lifetime in the solid, which is measured as the time difference between the creation of the 1.273-MeV γ-quantum and the annihilation radiation (two 511-keV γ rays). The natural lifetime of Ps of 142 ns is reduced by annihilation with electrons during collisions. The lifetimes, the inverse of the annihilation rates, become longer when a positron or positronium is localized at spaces with lower electron density such as voids.

Thus, positrons can be used as a probe to investigate average sizes of the free volume, size distribution, and the free volume concentration by measuring their lifetimes. The raw data of PALS are plots of the annihilation radiation signal as a function of time. In the case of defect concentration porosities that are not too high, there will be at least two or three lifetime components τ_i in the spectra, which are usually analyzed as a sum of exponentials after background subtraction. Furthermore, the spectra can be deconvoluted with CONTIN software. Basically, PALS allows the calculation of both the porosity and the pore size distribution. Since the determination of pore size distributions is based on the trapping of positrons in voids of varying size, PALS was reported to lead to a single, average lifetime in the case of interconnected pore systems.

The fundamental problem in the evaluation of PALS data lies with the exact relationship between the intensity of the long-lived components of a PALS spectrum and the concentration of the voids of a certain size, which is still a matter of intense research. For nonperiodic solids, a frequently applied model was developed by Tao and Eldrup. The extention of PALS to voids larger than 1 nm causes further uncertainties and is subject to ongoing research. PALS has been successfully applied to the investigation of porosity in various types of porous materials, such as thin siliceous films, cement, porous silicon, zeolites, and mesoporous silicas such as vycor glass.

Mercury Porosimetry

In mercury porosimetry (MP), gas is evacuated at elevated temperatures and low pressures from the sample placed inside a pressure chamber. Subsequently the sample is immersed in mercury and an external pressure is applied to force mercury into the sample. By gradually increasing the pressure and monitoring the incremental volume of mercury intruded for each new applied pressure, the pore size distribution of the sample can be estimated in terms of the volume of the pores intruded for a given diameter D. Mercury is a nonwetting fluid at room temperature for most porous materials of technological interest. The evaluation of pore sizes from MP is based on the Washburn equation, quantifying the pressure p required to force a nonwetting fluid into a circular cross-sectional capillary of diameter D by $p = (4\gamma \cos\theta)/D$ where γ is the surface tension of mercury and θ is the contact angle on the surface of a solid sample. MP allows the determination of average pore sizes (and their distribution) between 3 nm and 200 nm and is therefore inappropriate for microporous materials, but more suitable for pore sizes above ca. 30 nm than nitrogen sorption. As further parameters, the total pore volume V_{tot} is accessible from the total intruded volume of mercury at the highest pressure determined, and the total pore surface S is calculated from $S = 1/\gamma|\cos\theta| \int_0^{V_{tot}} p\,dV$. Because of inherent shortcuts MP is not used as frequently as gas absorption techniques for samples with mesopore sizes below 20 nm. During the measurement, high pressures used to force mercury into small pores may compress the sample. Damage or compression of highly porous silica has been reported previously. In addition, mercury porosimetry overestimates the volume of the smallest pores in the case of ink-bottle-shaped pores, because the intrusion of mercury into the larger pores is determined by the small openings. Moreover, it has to be pointed out that pore size distributions by the Washburn equation are not a geometrical relationship, but a physical characteristic of a porous medium, because MP is based on transport and relaxation phenomena.

General Strategies for Producing Porous Materials of Any Kind

The most successful way to produce all sorts of nanoporous materials is the templating method, shown schematically in figure. An organic (or sometimes inorganic) compound acts as a place holder what later becomes a void space in nanoporous materials. The templating concept obviously allows control, mainly of the criterion of pore size but also pore shape. At first, a suitable template structure has to be provided. By "suitable" we mean that this template structure must constantly be compatible with solvents involved in the process and the final network materials. Macrophase separation must be avoided at all times. As was pointed out previously, the creation of a porous material is always connected to the creation of a high surface area. In other words, the interface energy between the template structure and the network material despite this huge interface area has to be smaller than kT. Otherwise the material can rearrange into more stable, nonporous phases.

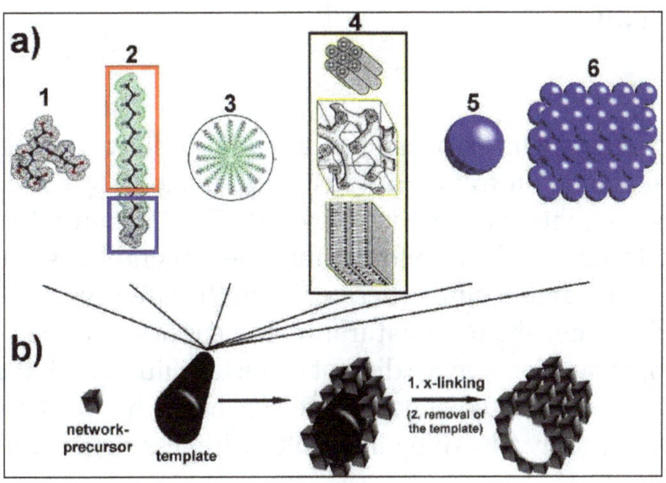

Schematic representation of the approach to obtaining nanoporous materials by templating strategies. (a) A range of applicable templates beginning with single molecules (1) or amphiphiles (2), followed by assemblies of amphiphiles as micelles (3) or lyotropic phases (4), followed by polymeric spheres (5) or assemblies of spheres (6). These units are surrounded or bonded to the network precursor systems in the actual templating step (b). The created pore resembles the size, shape, and even functional characteristics of the template in an ideal case. Finally, the template is removed. Therefore, it is possible to tune the properties and most importantly the size and interconnectivity of the pore and pore system in the order (1)–(6).

The template structure can have sizes on many different length scales. The smallest pores are achieved for molecules as templates. The pores created have a shape and size similar to those of the molecules used. The pores can be spatially uncorrelated with each other. In this case one speaks about molecular imprinting. Molecular imprinting is even successful for organic materials used to build networks and is therefore very often used for imprinting biomolecules to a matrix for biochemical applications. Highly ordered materials with pores that have a correlation with each other are found for zeolitic materials. The most prominent examples for zeolites are found in silica chemistry, and there are some naturally occurring compounds. Here it is enough to mention that primarily amines and ammonium cations as templates determine the formation of pores in the range of 1 nm to 1.5 nm. The materials have high symmetry and a crystalline framework.

Larger pores can now be achieved by assemblies of molecules forming templates. Here assemblies of amphiphiles are ideal because of their selforganizing properties and their capability for lowering interface energy. An amphiphilic molecule typically has one hydrophilic and one hydrophobic part, which tend to lead to microphase separation. The resulting template structures are micelles or ultimately lyotropic phases with different possible symmetries (lamellar, hexagonal, or gyroid). Materials with pores on the order of 2–50 nm (so called mesoporous materials) result, with a sometimes surprisingly high order. Alternatively hydrophilic polymers can be adopted as a template, which is not further considered in this article. Even larger pores are obtained when polymer spheres or colloidal crystals of these sphere are used as templates.

Inorganic Porous Materials

Microporous Materials

Zeolites

By definition, zeolites are crystalline hydrated tectoaluminosilicates of alkali and alkaline-earth cations with fully cross-linked open-framework structures made up by corner-sharing TO_4 tetrahedra, where T represents a tetrahedral atom (i.e., Si or Al). The first natural zeolite mineral (stilbite) was discovered in 1756 by Cronstedt, who found that the mineral loses water rapidly on heating and thus seems to boil. The name zeolite is derived from the Greek words zeo (to boil) and lithos (stone). Since Cronstedt's time, about 50 natural zeolite species have been discovered, but their potential for adsorption/separation science did not come to fruition until the twentieth century. In 1932, McBain recognized that activated or dehydrated chabazite has the property of functioning as a sieve on a molecular scale, only adsorbing molecules with less than certain critical dimensions.

In light of this selective sorption, he introduced the designation "molecular sieve" to describe this selective adsorption. A representative empirical formula of a zeolite is $M_2 / n\,O \cdot Al_2O_3 \cdot SiO_2 \cdot y\,H_2O$ where M represents an exchangeable cation of valence n.

M is generally a group I or II ion, although other metal, nonmetal, and organic cations may also balance the negative charge created by the presence of Al in the structure. The framework may contain cages and channels of discrete size, which are normally occupied by water. In addition to Si and Al, other elements can also be present in the zeolitic framework. They need not be isoelectronic with Si^{4+} or Al^{3+}, but must be able to occupy framework sites. Strictly speaking, only alumosilicates are referred to as zeolites. For example, the $AlPO_4$ framework consists of Al^{3+} and P^{5+} oxide tetrahedral in perfect alternation and therefore can be considered as the "3-5" analogue of the "4-4" pure SiO_2. However, $AlPO_4$'s, element-substituted $AlPO_4$'s, and other molecular sieves with framework T atoms other than silicon and aluminum are not zeolites.

Based on the pioneering work of Barrer, synthetic zeolites became available by the early 1950s. Researchers at the Linde company succeeded in synthesizing the zeolites A, X, and Y, which have become three of the most protable synthetic zeolites. Today, about 100 structurally different synthetic zeolites are known, and there are both synthetic analogues of natural zeolites and synthetic zeolites with no natural counterpart. In contrast to natural zeolites, synthetic zeolites may contain inorganic and organic cations, such as quaternary ammonium ions. Meanwhile, in total about 80 different molecular sieve structures are known, which are classified by the Structure Commission of the International Zeolite Association.

The framework of zeolites and analogues is based on an extensive 3D network in which the tetrahedral sites as the "primary building unit" are linked by oxygen atoms. The topology of the framework can be described in terms of a finite number of specific combinations of tetrahedral sites called "secondary building units" (SBUs). Zeolite frameworks are built up by one SBU type only. The framework topology can be considered to be made up by "tertiary" building units corresponding to different arrangements of the SBUs in space. For example, sodalite, zeolite A, and zeolite Y can all be generated by the truncated octahedron known as the "beta cage". In essence, the SBUs join to form zeolite channels known as oxygen windows that pass through the zeolite. It is the way in which these SBUs join together that gives rise to the huge number of different zeolites with intriguing properties. Although the aluminum content in zeolites can be equivalent to silicon, no Al-O-Al linkages have been observed in zeolites. Loewenstein attributed the absence of Al-O-Al linkages in tectoalumosilicates to the decreased stability of clusters of negative charges compared with isolated negative charges; therefore $Si/Al \geq 1$ for zeolites.

To provide a suitable classification of zeolite framework types, a topological description was introduced by Meier and Moeck, who established the concept of "coordination sequences," corresponding to the numbers of T atoms in the first, second, etc. tetrahedral coordination sphere of each topologically nonequivalent T site. The sequence of these numbers is characteristic for the different framework topologies and has turned out to be useful for determining the degree of similarity between different zeolites. Based on the IUPAC nomenclature for zeolites and analogues, framework types of different topologies are described by a three-letter code, representing all variants of a framework with a distinct topology, regardless of the chemical composition and cell dimension. For example, the sodalite structure is referred to as SOD and the zeolite Y structure as FAU.

In conclusion, the characteristic features of zeolites and analogues are a high surface area (about 1000 m^2/g), the highly ordered micropore channel system with uniform pore sizes and shape, and the presence of charges and ions at well-defined sites in the framework. The well-defined pore system results in the molecular sieving property, excluding molecules with a too large kinetic diameter from the pore system. The presence and the particular distribution of the charges and ions create a specific electric field within the framework, which permits the control of hydrophilicity and the introduction of ion-exchange properties.

Zeolites have found widespread applications as dehydrating agents, selective adsorbents, ion exchangers, and, most importantly, catalysts of both high activity and selectivity for a huge number of different reactions. Among various other applications, zeolites are used for the drying of refrigerants, removal of atmospheric pollutants such as SO_2, cryopumping, separation of N_2 and O_2, separation of paraffin hydrocarbons, recovery of radioactive ions from waste solutions, catalysis of hydrocarbon reactions, and curing of plastics and rubber. Certain zeolites show a strong Bronsted acidity.

Molecular sieves are selective, high-capacity adsorbents because of their high inner surface area and their strong interactions with adsorbates. The separation of molecules occurs on the basis of their size and shape relative to the size and geometry of the apertures of the sieve. Furthermore, the polarity of the adsorbates is a crucial parameter: for example, the separation of N_2 and O_2 in air on zeolite A, due to the fact that N_2 has a quadrupole moment different from zero.

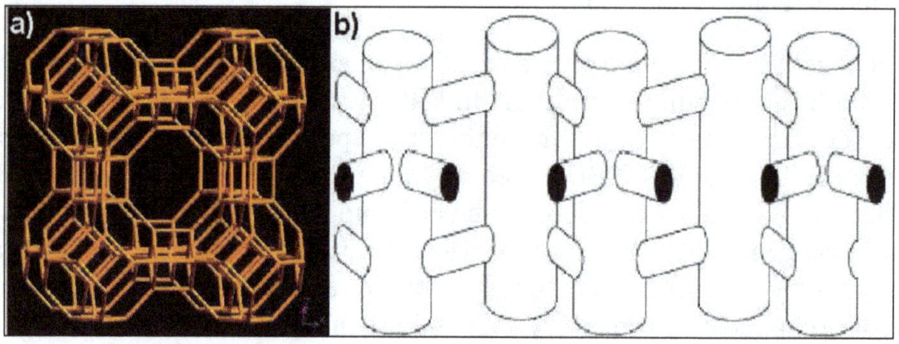

Two interesting zeolite cage systems. (a) LTA. (b) ZSM-5.

The most important application of molecular sieves is their use as catalysts, because zeolites combine a high acidity with shape selectivity, high surface area, and high thermal stability. Based on these properties, zeolites are used to catalyze a huge variety of hydrocarbon reactions, such as cracking, hydrocracking, alkylation, and isomerization. In these reactions, the reactivity and selectivity of the zeolites are determined by the charge imbalance between the silicon and aluminum atoms. One of the most famous examples of the industrial application of zeolites as catalysts is the isomerization of *o*-xylene to *p*-xylene, which is achieved by the use of ZSM-5, a zeolite with orthorhombic symmetry and an intersecting 2D pore structure. The zeolites of the ZSM-5 family have a high Si/Al ratio and two types of pores, both formed by 10-membered oxygen rings. The first of these pores is straight and elliptical in cross section; the second pores intersect the straight pores at right angles, in a zigzag pattern, and are circular in cross section. The size of these pores allows only the diffusion of molecules of a certain size through the channel system. While the conversion of *o*-xylene into *p*-xylene is usually accompanied by transalkylation reaction, yielding toluene and trimethylbenzene, a high isomerization-to-disproportion ratio is observed with ZSM-5 catalysts.

This selectivity is due to the fact that the pore size matches the diameter of xylene molecules. Moreover, ZSM-5 zeolites are widely used for MTG (methanol-togasoline) processes. Although the mechanisms are not fully understood in detail, *ab inito* calculations have started to shed some light on these processes.

A wide variety of processes have been developed for the synthesis of zeolites, which is described in detail by Davis and Lobo. In essence, the zeolite syntheses, introduced in the 1940s, involve a hydrothermal process in the presence of a large excess of water and the use of reactive alkali/alkaline-earth aluminosilicate gels at comparably low temperatures (about 100 °C) and autogeneous pressures. This synthetic pathway is supposed to be based on the structure-directing effect of hydrated alkali-metal cations, controlling the organization of zeolite subunits and a solution-mediated crystallization of the amorphous gel. The use of alkylammonium cations in the zeolite synthesis represented a significant improvement. Organic cations such as tetramethylammonium (TMA^+) made it possible to achieve a significant increase in the Si/Al ratio in zeolites such as A, X, and Y. Since the pioneering work of Barrer, the use of organic materials in zeolite syntheses has been strongly expanded, and new zeolite structures have been obtained. As representative examples, ZSM-5 and ZSM-11 are synthesized at high pH (usually >10) with the use of tetrapropylammonium and tetrabutylammonium cations, respectively.

A further major step in the synthesis of molecular sieves was the synthesis of aluminophosphate molecular sieves ($AlPO_4$) by researcher. Meanwhile, $AlPO_{4-5}$ can be synthezied with more than 20 different organic amines, for example, quaternary ammonium cations, diamines, cyclic amines, etc. The only aluminophosphate that can be prepared without organic structure-directing molecules is VPI-5. In contrast to zeolites, the synthesis of $AlPO_4$ is carried out under acidic or only slightly basic conditions. Since 1982 several new types of molecular sieves, based on $AlPO_4$, have been synthesized, including aluminophosphates, silicoaluminophosphates, metalloaluminophosphates, and metallosilico-aluminophosphates. Several applications have been developed for aluminophosphate-based molecular sieves, for example, as hosts for laser dyes, photoswitchable dyes and fluorescent dyes, catalysts, and hosts for chemical sensors.

In spite of these advances in the synthesis of zeolites, it is not possible to "design" zeolites in the true sense of tailoring a certain recipe with regard to predefined desired properties of the final molecular sieve. Furthermore, the details of the mechanism of zeolite synthesis are still a matter of discussion. In principle, two pathways have been proposed for the zeolite synthesis mechanism. In the "solid-solid transformation mechanism," crystallization occurs directly from the amorphous gel through a reorganization of the gel. Researchers were able to establish a nonaqueous synthesis of ZSM-5, which was shown to follow a solid-phase transformation mechanism. This mechanism for ZSM-5 was confirmed in further studies. In the "solution crystallization" or "solution-mediated transport" mechanism, small species such as aluminate, silicate or aluminosilicate ions diffuse from the liquid phase to nucleation sites of crystal growth. It is assumed that an equilibrium exists between the solid-gel phase and the solution and that the nucleation takes place in the solution. The gel is supposed to dissolve, and the dissolved entities diffuse to the nucleation sites in the solution. Several studies indicate that some zeolites can be synthesized by either of the mechanisms and that zeolites can also be crystallized from a single-solution system without gel phase. It was also proposed that in certain cases the two types of mechanisms may occur simultaneously.

Since most of the molecular sieves are prepared under nonequilibrium conditions, zeolites are metastable systems. The synthesis of aluminum-rich zeolites from aluminosilicate gels follows Ostwald's law of successive transformations, which states that a metastable phase is replaced by a thermodynamically more stable phase, etc., until the most stable phase is developed. However, this rule does not apply if organic cations such as TMA^+ are added, which may not lead to the formation of the thermodynamically most stable zeolites. Moreover, the role of organic molecules in creating the uniform cage and pore structure is not fully understood, particularly regarding their function as "space filler," "structure-directing agent," or a "template" in the true sense during the gelation or nucleation step, which is discussed in detail by Davis and Lobo. For example, the fact that at least 22 different organic molecules can be used to synthesize ZSM-5 shows that in this case the organic molecules act as space fillers rather than as a structure-directing agent. So far, only one case is known where an organic molecule acts as a true template.

Since a detailed description of the current research is beyond the scope of this overview, only a few interesting novelties in this field can be presented here. A great deal of work has been attributed to the development of zeolites with larger pore sizes. The aluminophosphate VPI-5 represents a substantial improvement in this direction. In general, large pores can be realized if the cages contain rings with more than 12 T-group oxygens making up the pore. VPI-5 has circular 1D channels with a pore diameter of 1.2 nm, which signi. cantly exceeds classical zeolites such as zeolite A or Y. Aside from VPI-5, various other large-pore aluminophosphates, metal phosphates, and high-silica zeolite structures have been reported. A further important improvementin zeolite science is the incorporation of zeolites into hierarchical structures. Meanwhile, molecular sieves can be prepared as hollow spheres, tubes and fibers, freestanding films, supported films, and thin films and layers. Oriented zeolite films were also reported recently. The application of zeolites as low dielectric films represents a further topic of industrial significance.

An important example of progress in the field of crystalline microporous materials is the development of "metal-organic frameworks" (MOFs), introduced by Yaghi and co-workers. These materials are made up by the coordination of metal ions to organic linkers, thus creating highly porous metal-organic architectures with substantial thermal stability up to 300 °C. MOFs have the lowest density of all crystalline materials. Although MOFs can still not compete with zeolites in terms of mechanical and thermal stability because of the large content of organic moieties, several interesting niche applications can envisioned. For instance, MOFs were reported to show catalytic activity and to have a high methane storage capacity.

Carbon

Carbon is found in nature in its chemical allotropes of graphite and diamond and occurs in a large variety of man-made species such as activated carbon, glassy carbon, carbon fibers, etc., which have been developed for a huge diversity of applications. In spite of the large number of different materials, microstructures of most of the porous carbons are derived from the graphite structure. Graphite is a crystalline material and consists of a succession of layers parallel to the basal plane of hexagonally linked carbon atoms. In this lattice, the interatomic distance within a layer is 0.1415 nm and the interlayer distance between the planes is 0.3354 nm.

In contrast to graphite, "nongraphitic" carbons are characterized by a random translational and rotational disorder of the planes ("graphenes"). The matrix structure of these materials is based on graphenes of finite size L_a and graphene stacks of finite height L_c, which possess a pronounced variance in the interlayer spacing *d*. Carbons with this particular kind of structural disorder are also called "turbostratic." Moreover, in turbostratic carbons the graphene layers show a significant variability in their shape and may also be bent. Nongraphitic porous carbons such as activated carbons contain a twisted network of defective graphene layers, cross-linked by aliphatic bridging groups. The substantial porosity in these materials arises from the imperfect packing and bending of the layers; the imperfect space filling of this packing results in microvoids/micropores of needle-like shape. It was previously shown that the boundaries of the micropores are mainly formed by the at surfaces of the layers.

The structure of porous carbons on the nanometer scale was elucidated by X-ray diffraction techniques and electron microscopy techniques. While electron microscopy studies of nongraphitic carbons suffer from both the low contrast of carbon and considerable disorder in the structure, SAXS and WAXS have provided deeper insights into the graphene structure and porosity. In particular, Ruland and co-workers have carried out detailed SAXS studies on various types of microporous carbons and described the changes in the porosity upon certain treatment procedures. In these studies, the concept of the "chord-length distribution" has turned out to be an appropriate quantitative characterization of carbonaceous pore structures possessing a pronounced dispersity in the pore size and shape. Whereas several SAXS studies claimed that certain porous carbons contain a fractal pore structure, recent detailed SAXS studies indicate that this apparent fractality is due to a misinterpretation of SAXS data.

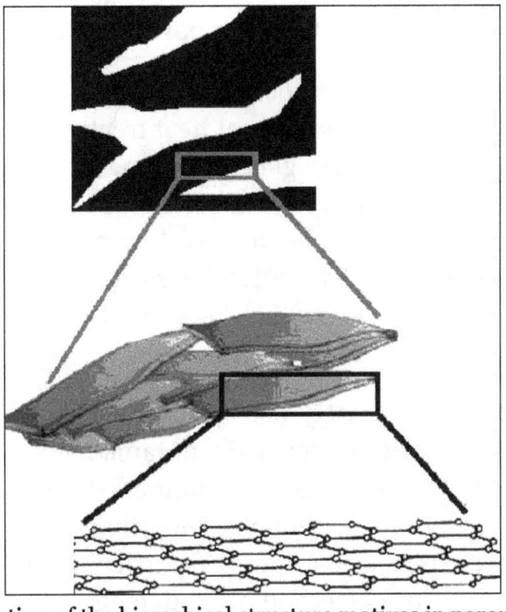

Representation of the hierachical structure motives in porous carbons.

The graphene stacks lead to diffuse maxima in WAXS curves, which are due to the translational and rotational disorder. With the application of suitable evaluation techniques, these data can provide a variety of structural parameters describing the size and disorder of the graphenes and the graphene stacks, particularly those data based on a recently reported improved method.

Certain materials such as tar pitch can be converted into crystalline graphite ("graphitizable carbons") by appropriate temperature treatments up to 2600 °C. The most common synthetic graphites are usually manufactured as composites in which ground petroleum coke is mixed into a paste with a coal tar pitch and then heat treated to around 1200–1400 °C in a calcining step to coke the pitch and drive all volatile material from the petroleum coke. Further heating to 2500–3000 °C causes an ordering of the carbon atoms to graphitize the mixture into true graphite. For other precursors such as poly (acrylnitrile) (PAN), almost no complete graphitization is possible, even at the highest possible treatment temperature.

A large variety of organic materials, such as wood, coconut shells, coal, synthetic polymers such as (PAN) and phenolic resins, and small organic molecules such as furfuryl alcohol, produce porous carbons upon suitable carbonization and activation processes. Among the most prominent porous carbons are charcoals, carbon blacks, carbon fibers, glassy carbons, and various types of activated carbons.

Charcoals are obtained from various sources such as wood, coconut shells, etc., and probably represent the class of porous carbons that has been used most extensively through the history of mankind. The medicinal use of charcoal is recorded as early as 1550 BC by the ancient Egyptians, though it was only in the latter part of the eighteenth century that its adsorbent powers were scientifically studied and it became possible to prepare it in a pure state, primarily as an antidote to poisoning. However, it does have many uses beyond the medicinal. It is found in the purification system of our drinking water, in air conditioning systems to remove odors, in linings for gas masks, and in nuclear power stations, where the charcoal adsorbs and holds radioactive contamination until the isotope decays. A variety of different activation processes have been developed to optimize the surface area (up to 2500 m^2/g) and the pore structure.

Glassy carbons, also known as "vitreous" or "polymer" carbon, are obtained by thermal degradation of phenolic resins under an inert atmosphere at heat treatment temperatures between 600 °C and 3000 °C. The name is derived from the glass-like appearance. Glassy carbon contains inaccessible nanometer-sized voids, which are also separated from each other. Because of this inaccessibility of the voids, GC is almost impermeable to liquids and gases, and the pore structure could only be investigated by SAXS techniques. This closed porosity can be opened by suitable activation procedures and creates a material with a sandwich-like structure, which can be used in electronics devices such as power density double-layer capacitors.

Carbon black represents a class of intensely black, finely divided forms of nongraphitic carbon with specific surface areas of up to 500 m^2/g usually obtained as soot from partial combustion of hydrocarbons. It is used as a reinforcing agent in automobile tires and other rubber products, but also as extremely black pigment of high hiding power in printing ink, paint, and carbon paper. The applications are dependent on the physical properties, particularly the size of the primary particles and the surface area. The most common method of producing carbon black is the furnace process. Briefly, this involves the partial combustion of a carbon-based fuel under rich conditions, resulting in the production of soot or carbon black.

The designation "activated carbon" does not refer to a particular material or morphology, but is related to activation processes applied to precursor materials. Usually activated carbons are produced from natural materials (wood, coconut shells, etc.), coal, or polymers by a two-step process involving carbonization and activation. The surface area, dimensions, and distribution of the pores depend on the precursor and on the condition of carbonization and activation. Various physical activation treatments have been developed, such as heat treatment in an inert atmosphere (up to about 900 °C) or in a water vapor atmosphere ("steam activation"). In chemical activation processes the carbon precursor material is mixed with certain chemicals such as zinc chloride, phosphoric acid, barium chloride, or KOH and then heated in an inert atmosphere up to temperatures of 1000 °C.

The exact mechanisms of the chemical treatment are still unclear; it has been suggested that the chemical activation involves dehydration of the organic material, leading to carbonization. SAXS studies have shed some more light on the changes in the microstructure of microporous nongraphitic carbons upon temperature treatment. In essence, the thermal treatment leads to an improvement of the packing of the graphene stacks in terms of a decrease in the variance of the d-spacing and an increase in the average number of graphene layers per stack. Simultaneously, the graphene size L_a also increases. Based on carefully performed SAXS studies, it was demonstrated that these changes in the matrix structure are accompanied by an increase in the average pore size and a substantial enlargement of the inner surface area. The latter phenomenon is the main reason for performing activation treatments and has led to the enormous industrial importance of activated carbons.

Aside from zeolites, activated carbons represent the class of porous materials with the highest industrial significance, partly because of their comparably low production costs. The most important applications are separation of gases, purification, and catalysis. Activated carbons are widely used to separate and clean gases, as in the separation of nitrogen and oxygen, and the separation of hydrogen in the recycling of off-gas from coke ovens and from ammonia. Furthermore, activated carbons are used to separate ole. ns from isoprene and butadiene and other hydrocarbon gases. The standard processes applied for gas separation are "pressure swing adsorption" and "vacuum swing adsorption".

Various applications of activated carbons have been developed in spite of certain drawbacks, such as the low inherent acidity compared with zeolites. As exemplary implementations, activated carbons are used for the isomerization of meta-xylene, as a co-catalyst for Fischer-Tropsch reactions, and for hydrogen sulfide oxidation. A recently established process provides the cyclic dimerization of 1,3-butadiene to produce vinyl cyclohexene in a Diels-Alder reaction.

A further type of important porous carbon is carbon fibers, which can be obtained from the pyrolysis of PAN. The structure has been investigated intensively in several SAXS studies; an extensive overview was presented by Ruland. Porous carbons can also be prepared from mesophase pitch, which is not described in further detail here.

While the basic features of the structure of porous carbons have been elucidated in the past 40 years, the current research in this field is focused on the details of pore structure and chemical composition. The different kinds of porous carbons still contain noncarbon elements, particularly hydrogen and oxygen, which, for instance, determine the inherent acidic and basic properties of activated carbons. The acidic behavior is associated with oxygen surface functionalities such as carboxyl, lactone, and phenol groups, which are located at the edges of the graphene layers. However, a wide range of oxygen surface functionalities can be envisioned, and a better understanding of their role in heterogeneous catalysis is needed. In recent studies, the attempt was made to get deeper insights into the nature of the oxygen species by various techniques such as UV-vis spectroscopy, inelastic neutron scattering, X-ray photoelectron spectroscopy, and titration methods.

Further progress has been achieved in the preparation of new types of microporous carbons. Recently, the synthesis of ordered microporous carbons was reported, based on the use of zeolites as a template. Although the zeolite structure is not completely preserved, it was shown that these microporous carbons possess a high surface area and exhibit a long-range micropore order, as proved by X-ray diffraction.

Although carbon nanotubes (CNs) may not be regarded as true porous carbon materials, it seems justified to include them in this class because of their inherent porosity provided by the hollow nanotube cylinders. Much interest has been attracted by CNs since their discovery. Two distinct types of CN have been produced: single-wall CNs (SWNTs) consist of a single graphite sheet wrapped into a cylindrical tube. Concentric multiwalled nanotubes (MWNTs) represent concentrically nested coaxial tubes ("Russian doll" structure). Furthermore, MWNTs can exist in the form of rolled-up carbon sheets (scroll structure). Potential applications of CNs have been reviewed recently by researcher.

Mesoporous Materials

As described before, mesoporous materials have pore sizes in the range between 2 nm and 50 nm. If a high porosity (larger than 50%) for a material can be achieved, porosity on the mesoscale is always equivalent to a high surface area. For instance, a material that has 80% spherical 5-nm voids can theoretically contribute a 480 m^2/g mesopore surface area. However, one very important differentiation has to be made: whether the pore system is ordered. The typical features of an ordered pore system are that the pores are monodisperse in size and have a specific shape and a mutual 3D correlation between the pores. When, on the other hand, a template above its percolation threshold has been applied, disordered pore systems result. Both of these classes of materials (ordered or disordered) are highly important for practical applications, as in sensing, catalysis, dielectric coatings, and, of course, typical molecular sieve applications. Next we focus mainly on materials with ordered pore systems because it is believed that the properties of these materials can be designed to a very high degree.

Silica

The highest degree of perfection in design has so far been achieved for ordered mesoporous silica materials. Ordered mesoporous silica materials can be obtained that have a variety of pore sizes (2–80 nm), a variety of pore shapes (from spheres to lamellae), and a variety of surface properties. This makes ordered mesoporous silica materials a very active field of research. There are three main reasons that have to be considered:

- Silica can be produced by sol-gel methods. The precursors used are not very reactive despite other precursors for oxidic materials such as titania, tin oxide, etc. This makes the silica sol-gel process very easy to control. Alkoxysilane compounds are used in most cases. The general process that is going on is given by the following reaction scheme:

  $$Si(OR)_4 \xrightarrow{H_2O} \left[Si(OH)_4\right] \rightarrow SiO_2$$

  Furthermore, there has been a great deal of investigation into the silica sol-gel process, and nearly every detail is known.

- The free energy of silica is very close to that of its most stable form, crystalline quartz. Furthermore, the Si-O bonds are so strong that the silica sol-gel process has a highly irreversible character. Amorphous silica is stable over a very high temperature range and period of time. This means that the shape of silica can be formed at room temperature and then be retained practically indefinitely. Once more, this distinguishes silica from other oxidic materials where crystallization frequently leads to a restructuring in the network and therefore can destroy the mesostructure.
- Finally, the interactions between surfactants and amphiphiles as surfactants or block copolymers can be tuned because silica units are available in anionic, neutral, or cationic form, depending on the pH. Furthermore, the charge density on these silica precursors matches the charge density of the surfactant headgroups quite well. Silica precursors and amphiphiles are compatible.

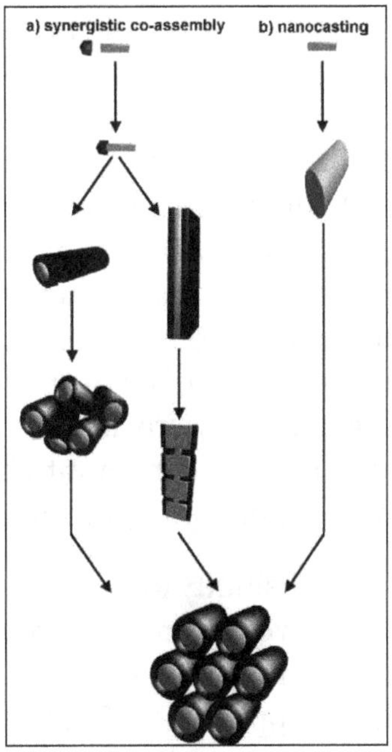

Schematic representation of the mechanisms involved in the formation of hexagonally ordered porous materials within the scope of templating strategies. One pathway is characterized by an interaction of amphiphile and network precursor, leading to the nanostructured composite phase (synergistic co-assembly). In the other case, a template structure is formed at first and imprinted precisely into a solid mold. Figure above shows the formation mechanism of ordered mesoporous silica materials. There are two different pathways: (a) the synergistic co-assembly mechanism and (b) the nanocasting or true-liquid crystal mechanism. The surfactant (hexadecytrimethylammonium bromide, CTAB, for instance) concentration that was used for the preparation of MCM-41 was way behind the critical micelle concentration and extremely far from the lyotropic phase regime. It was therefore quite surprising in the beginning that MCM-41, with its hexagonal pore system mimicking a hexagonal liquid crystal, had formed. It is accepted in the meantime that the first step in the formation of MCM-41 is an interaction between the cationic surfactant headgroup and anionic silica species. This special interaction between silica species and surfactant is generally denoted as S^+I. If one considers the packing parameter $P = \frac{V_{tail}}{l_{tail}\, a_{head}}$ where V_{tail} is the volume of the hydrophobic chain, l_{tail} is its length, and *a*head is the surface the headgroup stabilizes, one realizes that the bonding of the silica species will modify the headgroup region. In principle this creates a new, silica-organic hybrid surfactant with a different packing parameter and therefore a different phase behavior. Researcher argued via N-NMR spectroscopy that these new surfactants form cylindrical surfactant micelles that have at first no mutual correlation to each other. Then, by cross-linking of the individual micelles, ordering toward MCM-41 occurs. This mechanism was further supported by electron paramagnetic measurement (EPR) performed by researcher. The model that is currently favored describes the formation of a lamellar system at first. Researchers argued that then silica condensation occurs, which decreases the charge in the silica layers. Because the charge balance between the surfactant and the silica has to be equal at all times, the silica sheets bend farther and farther until just isolated cylinders in a hexagonal alignment remain, which is MCM-41. This process is known as charge density matching. When the recipe for the MCM-41 preparation is modified slightly, one can also obtain porous silica materials with different morphologies. The collection of these materials is called the M41S "family." There are MCM-41 (MCM = mobile composition of matter), MCM-48 with a gyroidal structure and MCM-50 with a lamellar structure. In the case of MCM-50 it is not possible to remove the template because the pore systems of lamellar compounds will break down. The structure of MCM-48 is most intriguing (it has two independent pore systems), and there has been much effort toward developing good synthesis methods for this particular compound. To obtain MCM-48 the surfactant-to-silica ratio should exceed 1; ethanol originating from the silica precursor also plays a crucial role. There are many other examples for the SCI*f* approach, and sometimes, quite unexpectedly, new structures are obtained. Researcher reported the formation of a L_3 sponge phase porous silica material when they used cetylpyrdinium chloride. Researcher were able to produce ordered mesoporous materials with pore symmetries not known in pure surfactant chemistry (for instance, space group *Pm*N3*n*) by employing so-called gemini or bolaform surfactants.

However, the S^+I pathway is not the only possibility for obtaining ordered mesoporous materials via the synergistic co-assembly mechanism. Every combination that results in an attraction between surfactant and silica source can be chosen in principle. There are also S^-X^+I (counter-ion-mediated) and $S^+X^-I^+$ (reverse charge matching) approaches. But much more important and widely used is the neutral S^oX^o approach using non-ionic template systems.

By adjusting the pH to 2 (the isoelectric point of silica) it is possible to obtain noncharged silica species. These species can interact with neutral templates (as non-ionic surfactant phases) via hydrogen bonding. One of the big advantages of using neutral templates is that the interaction with the inorganic network is weak enough that extraction techniques instead of calcination can be applied to obtain the porous materials. The first attempt to use neutral amphiphiles was amines with ethanol as a cosolvent, and except for some textural differences and a larger wall thickness, the MCM-41 structure type could be reproduced. A little later researchers reported the first examples of an ordered mesoporous silica material derived by alkyl (C_{11-15})-polyethyleneoxide (PEO) surfactants. In contrast to MCM-41, which gives the best ordered materials under hydrothermal conditions, these new materials can be obtained in good quality under ambient conditions.

Often, the formation of neutrally templated silica materials takes a pathway different from the co-assembly that is valid for the M41S materials. Researcher where the first ones to use a high-concentration phase of used poly(ethylene glycol) alkyl ethers like $C_{12/16}(EO)_8$ as a true liquid crystal template. The resulting pore systems were imprints of the preceding lyotropic liquid crystal. Because there is a 1:1 relation between the template phase and the pores as nicely shown for poly (butadiene)-poly(ethylene oxide) PB-PEO block copolymers, this process was also termed "nanocasting." While the synergistic co-assembly (MCM route) naturally leads to powder-like materials or, in the best cases, to larger particles, the nanocasting makes it possible to produce monoliths or thin films.

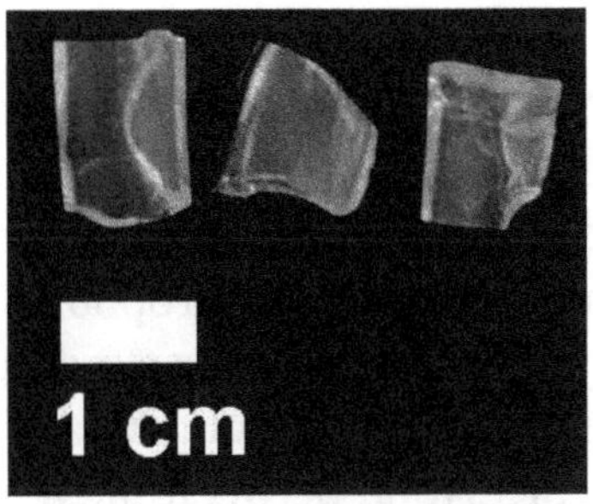

Photographic images of three nanoporous silica monoliths obtained by the nanocasting approach. These materials are truly nanoporous. The template (an amphilic block copolymer) was extracted by liquid-liquid extraction, and the solvent was then removed by supercritical drying.

This allows for the first time the division of a macroscopic volume into well-defined, nanometer-sized compartments. A crucial problem in maintaining the macroscopic and microscopic integrity of the materials is the removal of the template and drying of the samples. Calcination fails for the production of crack-free monolithic samples. Fortunately, the templates can be removed by liquid-liquid extraction. However, capillary stress due to evaporation can very easily lead to cracks. Capillary pressures generated by a phase transition of the pore-fluid to the gas state can be avoided by supercritical drying. Cracks can be further avoided by adding silica fibers to the sol used for the nanocasting process.

To extend the accessible pore sizes that could be achieved so far (≈5 nm), the range of nonionic amphiphiles was extended toward block copolymers. Here, poly (styrene)-PEO block copolymers and pluronic (PEO-poly (propylene oxide)-PEO) block copolymers as templates were used, with pore sizes reaching 10 nm. The nanocasting approach can also be used to produce materials with ordered mesopores of a maximum size around 80 nm when suitable block copolymers are available. Two nice examples of hexagonal pore systems obtained by block copolymer nanocasting are shown in figure.

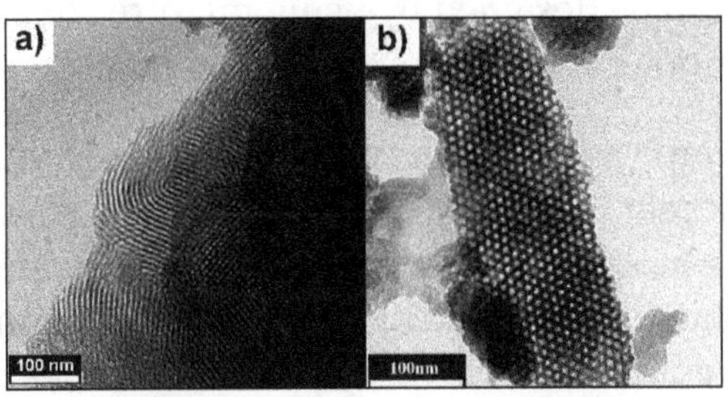

TEM images of two nice examples of hexagonally organized pore systems.

When these materials are analyzed by nitrogen sorption analysis it is seen that the total surface areas can reach values as high as 1000 m^2/g. Taking into account that a pure mesopore system can just contribute on the order of 400 m^2/g and the already high uptake of gas for low pressures, it has to be concluded that these materials are microporous as well. This was a point of discussion for a long time because it was clear that the additional microporosity will affect the properties of the materials significantly. The micropores can contribute as much as 80% to the total surface area in some cases. It is accepted in the meantime that the microporosity is caused by the PEO chain imbedded in the network matrix during nanocasting. The hydrophilic PEO domains of the template are dispersed in the water/silicic acid mixture during the casting process and remain there when the network solidifies.

When the materials are calcined, these imbedded, single polymer chains lead to micropores. The most comprehensive study so far on the nanocasting process and the creation of micropores was published by Polarz and Smarsly. A quantitative relation between the size of the mesopore D_c and the composition of the block copolymer used (length of hydrophobic and hydrophilic block) was derived. From the dependence on both block sizes it is seen that a classical picture that describes the lyotropic crystal by phase-separated hydrophobic and hydrophilic parts might not be suitable. A so-called three-phase situation describes the correlation between the template structure and the pore much more accurately. Only for relatively large hydrophobic blocks is all of the PEO needed to stabilize the interface, and then it creates microporosity. A fraction of PEO is always deposited on the hydrophobic core, which makes the actual size of the mesopore larger. It was quantitatively here that the size of a pore depends in principle just on the average number of monomer units in the amphiphile.

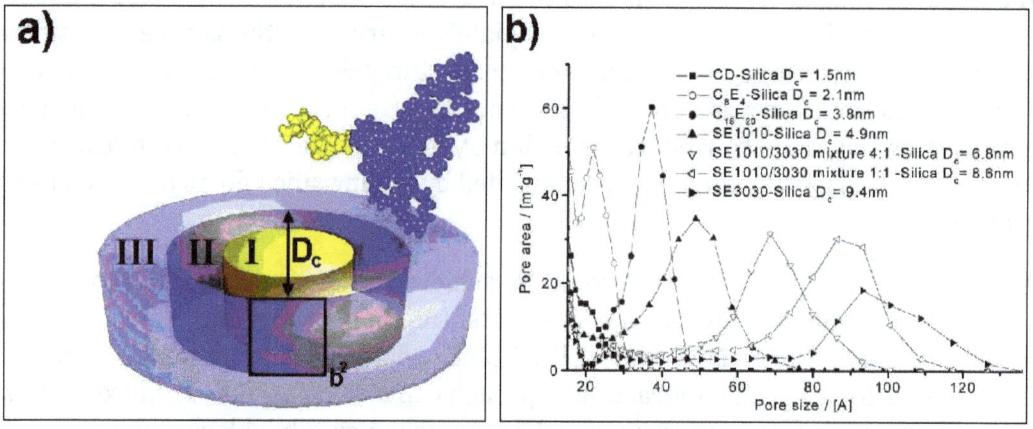

Structure of amphilic molecules in lyotropic structures acting as templates for mesopores.

The hydrophobic block is printed in yellow, the hydrophilic block in blue. (a) The three phases are depicted. (b) How pore sizes can be designed by using mixtures of amphiphiles as templates. The graph shows pore size distributions derived from sorption data.

This average number can be simply varied if one uses mixtures of amphiphiles (for instance a small and a large surfactant) in different amounts. In addition to the prediction of pore sizes from the molecular composition of pore sizes, it now became possible to achieve a fine-tuning of pore sizes by using mixtures of amphiphiles as templates. With the most commonly available amphiphiles (even pluronics) it is now possible to achieve every pore size between 1.5 and 10 nm.

There are other methods available to influence the pore size of ordered mesoporous materials. One way is to change the temperature during preparation of the material. The higher the temperature is, the larger the pores become; the electrolyte concentration is also an important factor. However, much more interesting is the use of hydrophobic swelling agents (such as trimethylbenzene). It should be noted that all of these factors (temperature, salts, and swelling agents) will change the packing parameters of the surfactant, which could lead to differences in phase behavior. An interesting case was reported by researcher when they added trimethylbenzene to the Pluronic P123 lyotropic liquid crystal phase, which normally leads to the hexagonally ordered SBA-15 material. The structure of SBA-15 is maintained only for low oil concentrations. At higher concentrations, the pores begin to buckle accompanied by an increase in pore size, and finally mesocellular foam is produced. One is actually able to follow the micelle-tomicroemulsion transition by looking at the casted, ordered mesoporous silica materials. This is a first indication that the nanocasting principle is capable of more than production of "designer materials"; it can also be used to investigate the structure of aqueous systems. The idea behind this is shown in figure.

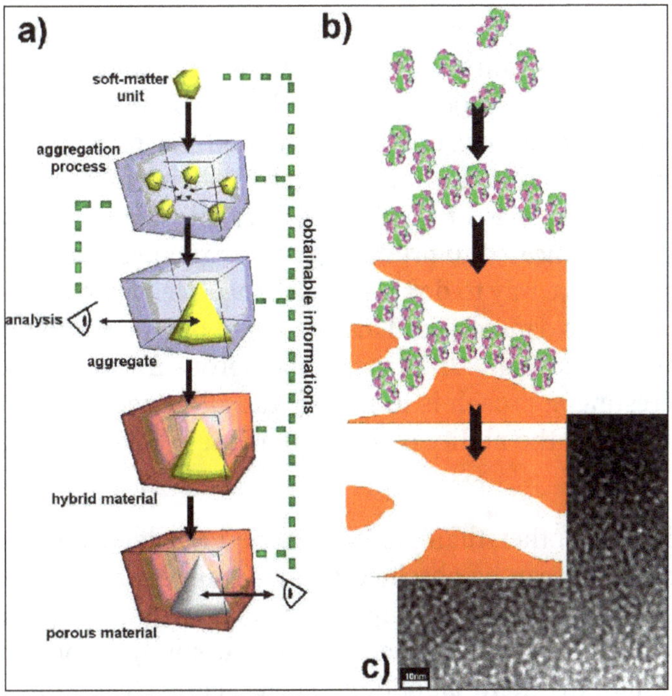

Schematic representation of the concept of using porous materials as a new analytical tool (a) and a practical example with cyclodextrins and their aggregation (b), which expresses itself in a porous material (c). When a soft matter unit is given in a solvent it is possible that aggregation into a more complex unit occurs. Via the nanocasting, which was described before, it is possible to catch these self-assembled aggregates and transform them into pores of the same size and shape. By looking at the pores one is able to evaluate the structure of the aggregate and to conclude steps in the self-assembly process. A nice example is given for cyclodextrins (b), which assemble to rod-like structures when placed in water. Accordingly, the pores (c) have a worm-hole architecture, which clearly indicates that the template was made from an assembly of cyclodextrin molecules.

A soft-matter (organic) unit such as a surfactant can aggregate when in water. Sometimes it is difficult to analyze this aggregate directly because the electron density does not differ much between solvent and aggregated structure. Techniques like electron microscopy fail (because of evaporation of the solvent under high vacuum), and even X-ray or light scattering give only limited information when the contrast is low. Furthermore, in a solvent the aggregated structures show rotational and translational motion, which also excludes techniques like atomic force microscopy. On the other hand, when a soft-matter aggregate is "nanocasted" into an inverted solid silica copy, all of these methods can be used, bringing all their benefits. In an ideal case one is able to gather information about each step shown in figure by investigating the pore structure of the materials. Researcher was able to bring more light into the assembly behavior of cyclodextrins in aqueous solutions by applying this concept. A solution of cyclodextrins was cast into a solid silica replica, and calcination led to the porous materials. It was shown that cyclodextrins assemble, in a manner similar to that of classical amphiphiles, into "worm-type" structures and that the driving force was a special kind of amphiphilicity the researchers called "quadrupolar amphiphiles." All of this information was obtained by studying the fine details of the pore structure with various analytical techniques (Fig shows one TEM image of the resulting porous materials).

The latter topics (microporosity in mesoporous materials and nanocasting as an analytical tool) already demonstrated how important it is, in order to determine the properties of a material, to complete a sufficient analysis. It can be summarized that so far ordered mesoporous materials can be obtained with a desired pore size, pore shape, and macroscopic structure. However, the composition of the network is limited to SiO_2, which is chemically quite inert. Concerning the reactivity and mechanical properties of these materials, this is a clear disadvantage. Therefore, there has been much effort dedicated to producing mesostructured materials with a composition different from that of silica. One way to do this is to change the network materials completely and go to other oxidic, more reactive materials. This is an area of its own that is discussed in Section. The second method is to deposit something inside the porous framework of the porous host, and the third and very interesting approach is to modify the silica network with organic groups. The latter two approaches still use the ability of silica to adopt even very small template features to form nanostructured materials.

Certain fractions of elements other than silicon can be imbedded into materials mostly of the MCM-41 type. These materials can be obtained when a source for this metal is added directly to the precursor sols. The most prominent example, which was already prepared in the original work by researcher, is Al-MCM-41, a MCM-41 where some Si atoms are replaced with Al. Through the incorporation of aluminum, the material obtains Lewis acidity. Therefore, Al-MCM-41 has been used for the catalytic cracking of various organic compounds. The performance of the competing zeolites is often better, although sometimes higher product selectivities with mesoporous materials were observed. Other catalytic materials with acidic functions are Ni-MCM-41 and Mo-MCM-41. These materials have some importance for the removal of sulfur from raw oil (hydro-desulfuration process, HDS). The latter cases (hydrocracking and HDS) are examples of why some major oil companies are even today quite interested in ordered mesoporous materials. The acidic forms of MCM-41 have also been used as classical organic synthesis catalysts, in Friedel-Crafts reactions and others. The great advantage of ordered mesoporous materials in comparison with zeolites is that MCM can also be used for large and bulky molecules, even when the reactivity is not so high.

Basic forms of MCM-41 have also been prepared, but here silica-organic hybrid materials are much more important, which will be discussed later. Cs-MCM-41 has some importance as a catalyst for Knoevenagel reactions. Another modified MCM-41, Ti-MCM-41 has been used for a variety of redox catalysis. Other examples of MCM-41 redox catalysts are Ti-, Fe-, Cr-, V-, B-, Al-, and Mn-MCM-41 materials, but Sn-, Zr-, and Ru-MCM-41 have also been invented. A different way to enrich silica MCM-41 with catalytic properties is to depositmetal or oxidic nanoparticles inside the mesoporous framework. It was shown recently that the size of the resulting nanoparticles can be fine-tuned by the size of the pore restriction and that the wall properties are of major significance.

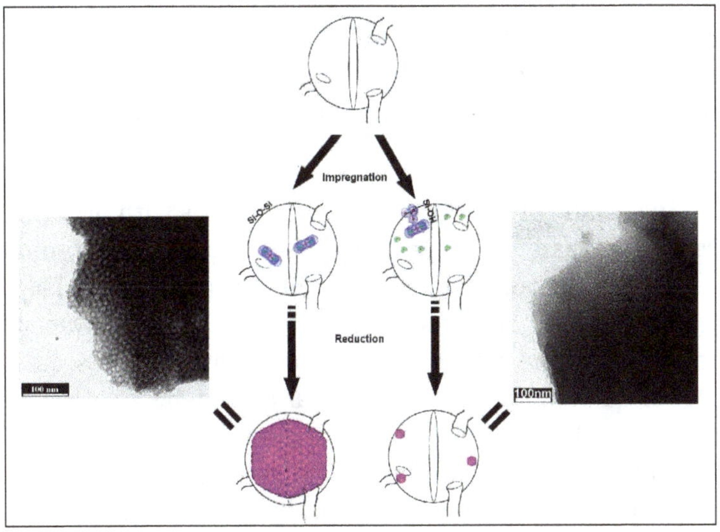

It is shown how the pore itself can be used to create new nanostructured matter, such as metal nanoparticles. When the interaction between the pore walls and the precursors is tuned it is possible to direct the growth of nanoparticles and to obtain control over their size and the properties of the hybrid materials. When the pore wall binds strongly to the metal precursors, Ostwald ripening is kinetically suppressed and subnanometer nanoparticles cover the pore walls. For repulsive or nonbonding interactions the size of the nanoparticles, their polydispersity, and their shape are dictated by the pore system.

As shown in figure, it is possible to create noble-metal particles that exactly resemble the size of the hosting mesopore after the reduction of suitable precursors. If the same process is applied to a material that is able to interact with these precursors (in the case shown a silanol-rich silica monolith interacts with $[PdCl_4]^{2-}$ ions) the created metal particles are bound to the pore surface and Ostwald ripening is suppressed.

Just subnanometer particles are created that are not visible in TEM images. One is therefore able to create ordered silica materials carrying a great variety of nanoparticles, semiconductor nanoparticles, or even oxide nanoparticles. Therefore much work has been devoted to preparing semiconductor nanoparticles by this route. Examples are gallium nitride, cadmium selenide, cadmium sulfide, silicon, titanium dioxide, germanium, gallium arsenide, zinc sulfide, and indium phosphide. Since then, preparation of colloidal particles inside mesoporous hosts has become a versatile tool for producing different kinds of nanostructures. One nice example can be found in the work of Huang, where silver nanowires were produced inside MCM-41. In this way, for example, metal nanowires have also been produced with Pd and Pt.

The last work indicates that it is possible to transform the whole pore system of a material into a new material that would not be accessible otherwise. The ordered mesopores act as a template again for new materials, which is released by removal of SiO_2 by HF etching. For instance, when an ordered mesoporous material is filled with a suitable precursor for carbon, depending on the pore morphology, particle-like carbon or nanoporous carbons (for instance, with the *Ia3d* structure of

MCM-48) have been prepared. Similar results can be obtained by preparing a replica of the pore structure with platinum. These results are extraordinary because now the structure of the new material resembles the structure of the organic liquid crystal that was used as a template for the production of the ordered mesoporous silica materials. Having an "artificial carbon liquid crystal" in hand could be very beneficial because carbon is stable to very high temperatures. The meso-structured carbon might therefore be taken as a new template for many materials that are not accessible with organic templates.

It should be mentioned that many of the host-guest compounds with ordered mesoporous materials are not of the inorganic-inorganic type. In many cases organic compounds as dyes or polymers have been immobilized in the porous framework preferentially of MCM-41. Polymers like poly-methyl-methacrcrylate were directly created inside the pores, and it was shown that the spatial con. nement had some impact on the physical properties of the polymers. Researcher reported a titanocene-containing MCM-41 material that was used as a solid-state catalyst for the preparation of a unique, high-molecular- weight poly (ethylene). Also very important in this respect are ordered mesoporous silica materials containing conductive polymers, as found, for instance, in the work of researcher for poly(aniline). The organic-inorganic hybrid materials created often have unique properties. One example of such unique features is new ordered mesoporous laser materials.

Generally, an active laser dye is brought into a porous environment, and its laser activity is studied. It was envisioned by researcher that a whole laser could be produced with the use of self-assembly laws. They could prepare a silica-surfactant-laser dye (rhodamine 6G) hybrid material with a fibrous morphology. The laser dye is located in the hydrophobic domains of the lyotropic template phase. The fiber was pumped perpendicular to the fiber direction, and above a certain threshold laser activity was detected. Researcher extended their work, and recently they prepared impressive waveguides by a combination of self-assembly and soft-lithography techniques. Mesoporous silica (template removed) was deposited on silicon and as the third layer there were the waveguides containing the dye-doped ordered mesoporous silica. The cylindrical channels were aligned throughout the whole waveguides, and even the dye seemed to be oriented in the template phase.

In all of the preceding cases the organic part of the material was just immobilized inside the pore system of the ordered mesoporous silica material. An alternative is to covalently link organic groups to the silica network and change the material properties this way. After calcination ordered mesoporous silica still possesses a certain amount of active silanol groups Si-OH. These groups can be reacted with a variety of organically modified sol-gel (Ormosil) precursors with the general formula $R\text{-}Si(OR')_3$. The attachment of a terminal organic group in a post-modification procedure leads to new, mesostructured silica-organic hybrids. However, the organic groups loom into the pore volume, and not every pore-surface Si group is modified in this way. These are two clear disadvantages of the post-functionalization method. In most cases amino functions have been introduced in this way into ordered mesoporous silica materials.

Materials containing ephidrine for asymmetric catalysis or organometallic complexes have also been invented. As was pointed out, it would be quite desirable if the pore wall itself were to be modified, instead of having the organic groups located in the pore volume. Therefore materials scientists tried with some success to add the Ormosil precursors directly to the synthesis mixture. This approach has two problems. The Si-O-R bond is not stable enough under most commonly needed synthesis conditions (high or low pH), and the organic group can negatively influence the interface energy, which is important when the amphiphile microphase separates and thus structures the material. Researcher could show that this pathway is successful for octyl and phenyl groups, and later for thiols, amines, epoxides, imidazoles, and other alkyl groups, when the Ormosil precursors with Si-C motives are used. The template can now just be removed by liquid-liquid extraction because of the presence of the organic groups. Therefore, and because of the effect of the organic ligand on the condensation rates, one obtains not fully condensed, quite weak, or collapsed materials. Still the amount of organic groups that can be incorporated in the materials remains quite low at 25%. If higher amounts of Ormosil precursor are taken, the structuring of the material via self-assembly is not successful.

A big improvement was made by researcher when they used bridging ligands (Si-R-Si) to modify the pore wall. These groups are truly incorporated into the material; they even seem to cover the pore walls. Now it was possible to use 100% of the organically modified sol-gel precursor for the synthesis, the materials are stable, and the pore walls are 100% covered with different organic groups such as ethylene or benzene.

It has been shown in this section that ordered mesoporous materials have become "designer" materials. They can be prepared with a great variety of properties such as pore size, pore shape, organic groups on pore walls, organic groups in pore volume, and any desired macroscopic shape of the materials. Next we considered whether the structural principles that work for silica can be extended to other inorganic network materials.

Transition Metals or Other Compositions

It was already suspected in 1993, briey after the invention of MCM-41, that one would be able to create mesoporous materials with compositions other than silica by surfactant templating routes because similar interactions between the network precursor and the amphiphile might be created. However, using network sources other than silica always involves two general problems. The precursors for these networks are much more reactive, which makes the whole process less controllable. Second, most other networks tend to exist in the crystallized rather than the amorphous form. The crystallization can result in major restructuring processes in the material and therefore might destroy the mesostructure. Most frameworks begin to crystallize at higher temperatures, when the surfactant is removed by calcination. This often makes it dif. cult to remove the template.

The first nonsiliceous examples of comparable materials followed in 1994; these were tungsten oxide, molybdenum oxide, antimony oxide, and vanadium oxiderelated materials, but it was not possible at that time to remove the organic template because lamellar structures were obtained. Researchers were the first ones to obtain transition metal oxides with structures similar to those of MCM-41. For instance, a mesoporous titania (TiO_2) material was obtained when the reactivity of the precursors was decreased by the addition of a complexing ligand as acac. Later, nicely structured niobium oxide and tantalum oxide materials were obtained by tuning the interaction of the amphiphile with the precursors (ligand-assisted route) for the inorganic framework.

Transition Metal Oxides

Because of their high relevance in many areas of catalysis and their variable redox and magnetic properties, much work was devoted to the creation of stable ordered mesoporous transition metal oxides. In the meantime, many compositions with Ti, Zr, V, Nb, Ta, Mo, W, Mn, and Y as the central element have been introduced. In comparison, most of the work has focused on ordered mesoporous titania because of its potential use in solar cells and electrochromic devices. Ordered mesoporous titania is a quite representative example for transition metal-based materials. The reactivity of titania precursors goes in the order $TiCl_4$ > $Ti(iOPr)_4$ > $Ti(OBu)_4$ > $Ti(OEt)_4$. These precursors all react rapidly with moisture and form TiO_2. To control the growth of a titania network around a template phase the reactivity has to be decreased. The most frequently used approach is to take $TiCl_4$ as a titania source and react it with anhydrous ethanolic solutions of the template (surfactant or block copolymer). The reaction between $TiCl_4$ and EtOH releases high amounts of HCl. High concentrations of HCl stabilize titania nanoclusters and therefore prevent uncontrolled condensation.

Moreover, one deals with an anhydrous solvent. The water needed for the self-organization of the amphiphile and for the condensation of the titania is provided exclusively by the atmospheric humidity. The humidity is the crucial point in these preparations, and it should be around 50%. It is also interesting to note that the water seems not to be able to diffuse from the top of a deposited film into the material, presumably because of densely packed surfactant hydrophobic chains on this surface. Diffusion of water and ordering of the material begin at cracks and propagate from there. As prepared, the materials have an amorphous framework that is often not desired for many applications. Crystallization of titania occurs at temperatures above 400 °C, which might also disrupt the mesostructure and leads to just 3-nm crystallites. As a possible solution titania nanocrystals were proposed as the starting material for ordered mesoporous titania as well.

- Metal Sulfides: Researcher used a true-liquid crystal approach to prepare various mesostructured metal sulfide materials. A lyotropic block copolymer phase is swollen by inorganic salt solution as Cd^{2+} salts, and then treatment with H_2S vapor follows. A nanocrystalline, ordered mesoporous CdS material is the result. Soon, a ZnS material followed. Similar to work where titania nanoclusters have been used for the preparation of mesotructured titania, several groups prepared ordered mesostructured germanium sulfide materials with Ge_4S_{10} clusters in the initial state.
- *$AlPO_4$*: Because of the similarity between silica molecular sieves (zeolites) and aluminum phosphates (ALPOs) there have been many attempts to synthesize MCM-41 analogous ALPO materials. In most of these cases non-stable materials are obtained or the pore systems are quite different from MCM-41. Researcher reported a nicely ordered hexagonal ALPO material in 1999, and researcher reported an interesting approach using an Al_{13} Keggin species as the source for aluminum.
- Silicon Nitride: Just as in ALPOs, one could suspect that it will be possible to produce an ordered mesostructured silicon nitride because the NH unit as in imides or amines is isoelectronic to oxygen, which would allow sol-gel chemistry similar to that of silica. There have been attempts to synthesize these compounds with silicon amines, ammonia, and nonaqueous solvents and surfactants.
- Al_2O_3: Ordered mesoporous alumina would also be a very interesting material because of its high relevance in catalysis. The classic way of preparing mesoporous alumina is by electrolytic etching (anodized alumina), and the results obtained are impressive for the

production of thin films. This technique is restricted to surfaces. Therefore, preparing ordered mesoporous alumina via templating routes should be very promising. It was found that it is very dif. cult to achieve highly ordered materials and to achieve control over pore size with alumina, although some examples exist. It seems that the S^oX^o pathway is appropriate for alumina.

- Metals: Researcher succeeded in the preparation of hexagonally ordered metal sponges via templating techniques. This technique is very similar to the described approaches, although the network is now formed by the reduction of noble-metal precursors.

Macroporous Materials

Opals and Inverted Opals

It is possible to cast inorganic materials by suitable templating techniques with high precision. Templates originating from the self-assembly of amphiphilic molecules have been discussed, and it was pointed out that colloidal objects or colloidal crystals can also be used as templates to obtain pores of a higher length scale. It is known that colloidal particles such as silica spheres, latex spheres, or inorganic crystals (such as gold or semiconductors), as soon as their size is fairly monodisperse, tend to pack into ordered assemblies called colloidal crystals.

SEM image of a nanoporous material formed by a colloidal crystal of latex spheres.

Figure shows the colloidal crystal formed by negatively charged 280-nm latex spheres. As can be seen from this image, the dense packing of 280-nm spheres creates regular voids in the size range of 70 nm. Therefore, colloidal crystals obtained by spheres or sphere templating techniques belong to the family of nanoporous materials. Methods for obtaining silica spheres can be found in the literature, and these rely mainly on the process invented by Stoeber. Latex spheres are mainly obtained by emulsion polymerization. However, common techniques for obtaining colloidal crystals of spheres are centrifugation, electrophoresis, and evaporation methods. Concerning the last method, the work of researcher, which is the convective self-assembly process, should be mentioned explicitly because it is a very good and easy method for obtaining opal structures of spheres. Very high ordered opal structures can be obtained, but the symmetry seems to be restricted to face-centered cubic (fcc), as expected from theoretical calculations. However, in rare cases other structures, such as (bcc), (hcp), and (rchp), could be observed. The spheres in the opal structures discussed so far are held together just by relatively weak forces. For later applications cracking or disrupting the structure is always a problem. By sintering or thermal annealing of silica spheres, or heating above the glass point for organic polymer spheres, it is possible to strengthen the material.

Much attention has recently been paid to the transformation of these opal structures into inverted opals by templating techniques. Because in these cases voids, sometimes on the order of several micrometers, are produced, just the basic principles are described in this review about nanoporous materials. A more detailed description can be found elsewhere. The interstices of the opals are filled with liquid or gaseous (CVD) precursors, which react to form a new network. Afterward, the sphere template is removed by calcination for polymer spheres or by HF etching for silica spheres. In this way, a large variety of inverted opals ranging from inorganic oxides such as titania, to metals, carbon, or semiconductors have been prepared.

Photonic Properties

All of the opal materials discussed show, because of their architecture, a characteristic modulation of dielectric constant in 3D space. This is the major requirement for so-called photonic materials. In simple terms, a modulation of the dielectric constant or the order of light leads to Bragg diffraction for specific light frequencies. This means that certain light frequencies cannot propagate through the opals. These frequencies where light is reflected are also called stop-bands. Further requirements for photonic materials are a high refractive index contrast and a high order in the material. Although the combination of Bragg's law and Snell's law gives a good approximation for the peak position and the dependence on the lattice constant, filling fraction, and refractive index contrast, a more detailed description is required:

$$\lambda_{c(hkl)} = 2 \cdot d_{(hkl)} \cdot \sqrt{\langle\varepsilon\rangle - \sin^2\theta}$$

where $\lambda_{(hkl)}$ is the wavelength for the stop-band, $d_{(hkl)}$ is the lattice constant, $\langle\varepsilon\rangle$is the effective refractive index, and θ is the angle. It is believed that the flow of light can be controlled with photonic materials in a way similar to that in which a semiconductor controls the flow of electrons. Indeed, the theories behind the behavior of light or electrons in solids are somewhat similar because both systems can be described as waves.

Table: Opals, semiconductors for light.

Semiconductors and electrons	Opals and photons
3D distribution of atoms.	3D distribution of scattering centers.
Periodic electrostatic potential.	Periodic dielectric constant.
Description by Schrödinger equation (scalar).	Description by wave equation (vectorial).
$\frac{\bar{h}}{2m}\frac{d^2\psi}{dx^2} + (E - V)\psi = 0$	$\frac{d^2E}{dx^2} + (n\omega)^2 E = 0$
Fermions.	Bosons.
Creates bands with allowed electronic states (energies) electronic states (energies).	Creates bands with allowed electronic states (energies) electronic states (energies).

However, because electrons are fermions and photons are bosons, there are also important differences. The 3D modulation in refractive index has consequences for photons similar to those the 3D modulation of electrostatic potential in a solid material has for electrons. Depending on the propagation direction of light inside the material, bands of certain energies are created that allow light of these frequencies to propagate. But more important, for some energies there are no states in the material, which means that light of this energy is not allowed inside the material. It is reflected, which is detectable by UV/Vis spectroscopy.

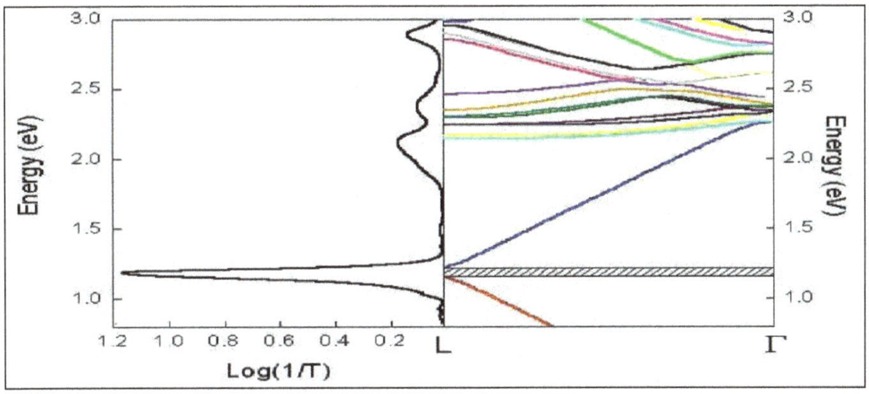

The band structure for allowed light states (right) is shown by the colored lines, depending on the propagation angle inside the opal (x axis). The left side of the figure shows the corresponding UV/vis spectrum.

Figure shows the correlation between the light-band structure inside a photonic crystal and the UV/vis spectrum obtained. As shown, there is a certain energy region (in this case above 1.0 eV) where no states for photons exist. As a consequence, no photons are transmitted through the material. Many, many challenges and exciting developments can be expected for this new area of nanoporous materials. Current interests focus on obtaining special structure types of opals, such as the diamond structure; improving the refractive index contrast; or preparing microengineered arrays of opals. Further applications of these materials lie in the areas of waveguides, filters, microcavity lasers, or optical chips.

Organic Porous Materials

The templating concept can be extended toward pure organic or polymeric compositions of materials. Similarly, an organic monomer is polymerized around a prefabricated template structure. The chemical and physical parameters of the polymer obtained differ very much from those of the monomer species if one compares with the sol-gel process of inorganic oxides. Often restructuring processes occur during polymerization that can lead to partial or complete phase separation. Therefore one has to distinguish:

- Transcriptive synthesis: The pore structure is a 1:1 imprint of the template structure.
- Reconstructive synthesis: The polymerization takes place under the spatial confinement imposed by the template, and morphological rearrangements lead to the final material.
- Synergistic synthesis: The monomer itself self-organizes.

Similar to all of the approaches discussed before, the interactions between the template phase and the network builder (or, in other words, charge density matching considerations) are of crucial importance. Furthermore, entropic forces such as restriction of polymer segment mobility due to spatial restrictions can become important. These factors decide whether a synthesis is transcriptive or reconstructive. One problem that also remains is that the porous materials are just metastable. In the case of polymeric materials this fact becomes much more important than for most inorganic networks. Because of swelling of polymers and their high flexibility, polymer systems might be unable to retain the porous structure in the long run.

One important field is the transcriptive translation of molecules into pores designed on a molecular level, which is also known as molecular imprinting. The synthesis pathway is similar to a key-lock or antibodyantigen mechanism. The template molecules (often aminoacids, sugars, peptides, or steroids) exhibit a certain pattern of distinct, spatially distributed functional groups. Suitable functional groups in the monomer systems interact with these patterns, and this leads to an imprint of the shape and functional structure. Finally, the template molecules are removed by extraction or chemical cleavage. Typical monomers that have been used are methacrylic acid and ethyleneglycol dimethacrylate. More detailed reviews of molecular imprinting used to obtain porous polymers can be found in the recent literature.

The second type for a clear transcriptive synthesis of porous organic polymer materials is found when colloidal crystals (opals) are used as templates. Macroporous structures are obtained. Possible polymer compositions involve polyurethane, poly(acrylate-methacrylates), poly(divinylbenzene)-copoly(ethyleneglycoldimethacrylate), polystyrene and poly(methyl methacrylate), and polypyrrole.

When micelles are used as templates a clear transcription to pores is questionable. Styrene/divinylbenzene polymers with spherical pores of 10–20 nm were obtained by templating with water-swollen AOT (sodium bis(2ethylhexyl)sulfosuccinate) micelles. The pore size appears to be too large to be caused even by swollen AOT micelles. In a similar work water-swollen AOT micelles were used for the preparation of ethylene trimethacrylate polymers in supercritical CO_2 with 20-nm pores. When supercritical CO_2 was omitted, pores of 100 nm were obtained, which already indicated the reconstructive character of the synthesis.

The micelles in the previous cases were already swollen with a polar solvent like water, while the continuous phase contained the hydrophobic monomer. The next logical step is to use microemulsions as templates for the production of porous polymers. Microemulsions are thermodynamically stable systems with droplet sizes in the range of 10–100 nm. Of particular interest for this application are bicontinuous microemulsions. It was seen in all reported cases that the pore morphology was no longer directly related to the structure of the microemulsion. Very often just latex particles were obtained. However, when hydrophilic (as acrylic acid) and hydrophobic (as methylmethacrylate) materials have been copolymerized in the presence of a microemulsion, true porous materials have been obtained. The final product was, as mentioned, very different, as expected from the morphology of the microemulsion. Therefore, one has to conclude that there is a reconstructive synthesis pathway. The transformation from a microemulsion to the final porous system for such a system was studied, and not less than four phase transitions have been observed during the polymerization reaction. The next level of complexity is reached when lyotropic phases are used as templates for porous polymers. First attempts to imprint a lyotropic structure into a polymer failed because of demixing. However, it was possible to get nicely ordered, reconstructed morphologies this way.

It can be concluded that a variety of nanoporous polymer materials can be obtained. But because of thermodynamic factors during the polymerization, demixing processes led to the formation of materials different from those planned. Nonequilibrium conditions might also play an important role. Although the prepared structures are very interesting and might have applications, it seems hard to synthesize designed nanoporous polymers. Furthermore, it seems that there are currently restrictions to a few polymer systems.

Nanocrystalline Materials

Nanocrystalline materials have been a hot research topic over the past 30 years. These materials abound in industry, bridging the gap between molecular and macroscale objects. Nanocrystalline materials are ultrafine-grained single-phase or multiphase polycrystals with grain sizes in the range of 1–100 nm, as depicted in Figure, the transmission electron microscopy (TEM) image of Fe-based nanocrystalline coating. In fact, the extremely small sizes and a large volume fraction of the atoms are located at the grain boundaries; on the other hand, these materials consist of about 50 vol.% crystalline component and 50 vol.% interfacial component.

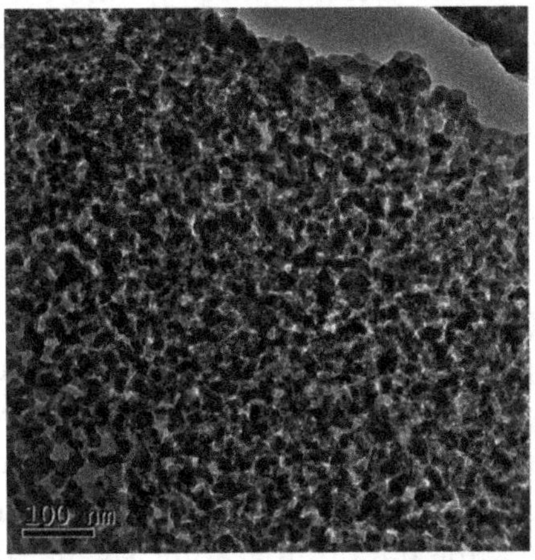

TEM image of Fe-based nanocrystalline coating.

It is recognizable that the nanocrystals are typically specified as anything, such as small grain polycrystalline materials, nanosynthesized surfaces, nanoparticles, and polymer micelles; each of them has varied usages, from drug delivery, to super capacitors, catalysts, and sensors. These materials are of interest for the following reasons:

- The properties of nanocrystalline materials differ from the properties of single crystals and coarse-grained polycrystals and are amorphous with the same chemical composition. This deviation is strongly related to the reduced crystallites size as well as the large amount of grain boundaries between adjacent crystallites.
- The concept of nanocrystalline materials seems to authorize the alloying of components which are immiscible in the solid or molten state. These fabricated alloys could be good candidates for advanced and technologically marvelous properties.

It is clearly seen that at a nanometric scale, the nanocrystalline materials contain a high grain boundary volume fraction; therefore grain boundaries and their interactions with crystal play a remarkable role in the different properties. It is important to point out that those nanocrystalline materials, as a new generation of advanced materials, have superior properties to conventional coarse-grained polycrystalline materials. They exhibit outstanding mechanical and physical properties such as high strength and hardness, low elastic modulus, improved ductility/toughness, excellent fatigue and wear resistance, increased diffusivity, higher electrical resistivity, reduced density, higher thermal expansion coefficient, enhanced specific heat, lower thermal conductivity, and better soft magnetic properties.

Nanocrystalline materials can be fabricated by gas condensation, plasma deposition, spray conversion technique, mechanical alloying, and some other methods. Obviously, there are two approaches to fabricate the nanocrystalline materials: "top-down" and "bottom-up." Both approaches play significant roles in industry and have some advantages and disadvantages. Bottom-up approach is nothing new in material synthesis. As a matter of fact, the typical synthesis of materials is to build atom by atom on a large scale and has been used for over a century in industrial applications. Bottom-up approach mentions the buildup of a material from the bottom as molecule by molecule, atom by atom, or cluster by cluster.

In the process of crystal growth, the growth species such as atoms, molecules, and ions assemble into crystal structure one after another after impinging onto the growth surface. Bottom-up approach also promises a preferable chance to get nanocrystalline materials with less defects, more homogeneous chemical composition, and higher short- and long-range ordering. It is recognizable that the bottom-up approach is driven mostly by the reduction of Gibbs free energy (ΔG), so that the nanocrystalline materials are in a state closer to a thermodynamic equilibrium state. In contrast to this, top-down approach most likely insets internal stress, in addition to contaminations and surface defects. Attrition or ball milling is a generic top-down method in making nanostructures, whereas the colloidal dispersion or gas-based reduction is a usual example of bottom-up approach. The former produces polycrystalline structures with irreproducible crystallography and poorly controlled grain orientation. The latter has generated a reproducible collection of structures and architectures, including alloys, pure metals, anisotropic nanostructures, and core-shell. In lithography, the process may be assumed as a hybrid approach, since the thin film growth is bottom-up and etching is top-down, while nanolithography is generally a bottom-up approach.

Nanocomposite Materials

Nanocomposites are materials that incorporate nanosized particles into a matrix of standard material. The result of the addition of nanoparticles is a drastic improvement in properties that can include mechanical strength, toughness and electrical or thermal conductivity. The effectiveness of the nanoparticles is such that the amount of material added is normally only between 0.5 and 5% by weight.

Improved Properties

Nanocomposites can dramatically improve properties like:

- Mechanical properties including strength, modulus and dimensional stability.
- Electrical conductivity.
- Decreased gas, water and hydrocarbon permeability.
- Flame retardancy.
- Thermal stability.

- Chemical resistance.
- Surface appearance.
- Optical clarity.

Nanoparticles have an extremely high surface to volume ratio which dramatically changes their properties when compared with their bulk sized equivalents. It also changes the way in which the nanoparticles bond with the bulk material. The result is that the composite can be many times improved with respect to the component parts. Some nanocomposite materials have been shown to be 1000 times tougher than the bulk component materials.

Applications

Nanocomposites are currently being used in a number of fields and new applications are being continuously developed. Applications for nanocomposites include:

- Thin-film capacitors for computer chip.
- Solid polymer electrolyes for batteries.
- Automotive engine parts and fuel tanks.
- Impellers and blades.
- Oxygen and gas barriers.
- Food packaging.

Types of Nanocomposites

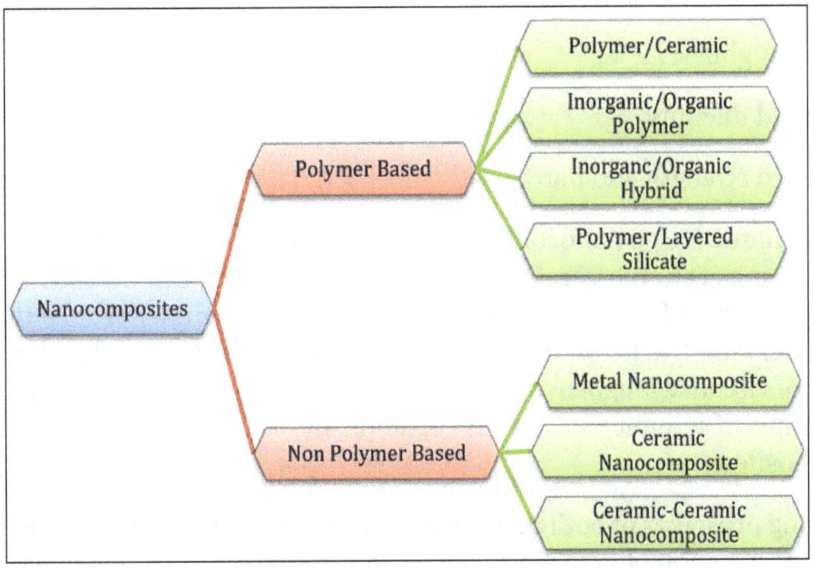

Classification of polymer- and non-polymer-based nanocomposites.

Nanocomposite materials can be classified in the following way based on the presence or absence of polymeric material in the composite. The nanocomposites in which the compositions do not contain any polymers or polymer-derived materials are called non-polymer-based nanocomposites. Non-polymer-based nanocomposites are also known as inorganic nanocomposites. They can be further classified into metal-based nanocomposites, ceramic-based nanocomposites, and ceramic-ceramic-based nanocomposites.

Non-polymer-based Nanocomposites

- Metal-based Nanocomposites: Bimetallic nanoparticles being investigated in detail in the form of either of alloy or core-shell structures due to their improved catalytic properties and advancement in optical properties related to individual and differentiate metals. They can be characterized by:
    - Super plasticity.
    - Lower melting points.
    - Increased strength and hardness.
    - Improved magnetic properties.
    - Increased electrical resistivity, etc.

  Non-polymer-based nanocomposites can be also classified as: metal/metal nanocomposites, for example Pt-Ru nanocomposites.
- Ceramic-based nanocomposites: Ceramic-based nanocomposites are defined as ceramic composites with more than one solid phase, in which at least one of the phases has dimensions in the nanoscale range (<50–100 nm). In these types of composites, both the phases have combined magnetic, chemical, optical, and mechanical properties, for example hydroxyapatite/titania nanocomposites. These can be characterized by:
    - Better toughness.
    - Increased ductility.
    - Increased strength and hardness.
- Ceramic-ceramic-based nanocomposites: The non-polymer-based nanocomposites can be also classified as ceramic/ceramic nanocomposites which can be used in the area of artificial joint implants for fracture failures and it could promptly reduce the cost of surgery and would extend the mobility of the patient. The life spam would increase by 30 years, if the use of zirconia-toughened alumina nanocomposite implants is used effectively. The other example of ceramic/ceramic nanocomposites are calcium sulfate-biomimetic apatite nanocomposites.

The most promising prospects of both metal-based nanocomposites and ceramic-based nanocomposites are in the application of areas in dentistry in which the non-polymer-based nanocomposites or the inorganic materials that is metal or ceramics such as calcium phosphate, hydroxyapatite, and bioactive glass nanoparticles are very advantageous in alveolar bone regeneration and enamel substitution.

References

- INTRODUCTION-TO-NANOMATERIALS-259118068: researchgate.net, Retrieved 10, August 2020
- Nanoparticle, science: britannica.com, Retrieved 16, March 2020
- Nanoporous-materials-27264580: researchgate.net, Retrieved 29, January 2020
- Nanocomposite-materials, nanotechnology-and-the-environment: intechopen.com, Retrieved 06, May 2020
- Introductory-chapter-nanocrystalline-materials: intechopen.com, Retrieved 24, July 2020

2

Nanostructure

The structures whose size lies between microscopic and molecular structures are known as nanostructures. Some of major examples of nanostructure are quantum dot, nanowire, nanofiber, nanorod, nanoring, nanomesh, nanoflower and nanotube. This chapter discusses in detail these nanostructures.

Quantum Dot

A quantum dot (QD) is an extremely small particle whose properties can be drastically changed merely by removing or adding an electron. In this parlance, an isolated atom can also be termed as a quantum dot. However, only a cluster of atoms or molecules demonstrates property variations. In biochemistry, QDs are often termed as redox groups. QDs are referred as quantum bits or qubits in nanotechnology. The dimensions of QDs are of the order of few nanometers. Nanotechnology is a multi-disciplinary field wherein the problem often includes studying biology, chemistry, computer science, electronics, etc. For example, consider a hypothetical biochip, which is grown analogous to a tree from a seed, and which might also include a computer inside it. In this analogy, both terms, a qubit or redox group, both are equally applicable. Besides, the classification of this chip as animate or inanimate, is difficult to accomplish. Each QD present in this chip may account for one or more data bits.

In some cases, an electron within the QD may occupy several states, thus, a QD may also represent a byte of data, instead of a bit (in case of qubit). Alternately, one QD might be simultaneously involved in more than one computations. In addition to these applications, other sophisticated uses of a QD may include nano-mechanics, neural networks, high density data storage, etc.

Working of Quantum Dots

Since quantum dots demonstrate size-dependent properties, their behavior can be easily tuned by controlling their size. Analogous to the atoms, quantum dots can be engineered to function as per requirements. For example, an atom can be excited by irradiating it with some energy: that is, an electron from a lower level can be excited to higher energy level. When this excited electron returns back to its original energy level, it emits light having energy equal to the energy difference between the two energy levels involved in the process. The color (or wavelength as well as frequency) of emitted light is characteristic to the atom. For instance, iron appears green when its atoms are excited upon exposure to heat, sodium gives off yellow color. This is attributed to the specific arrangement of energy levels within an atom. Thus, different atoms emit light of different colors or wavelength. This difference comes due to the discretization of energy levels.

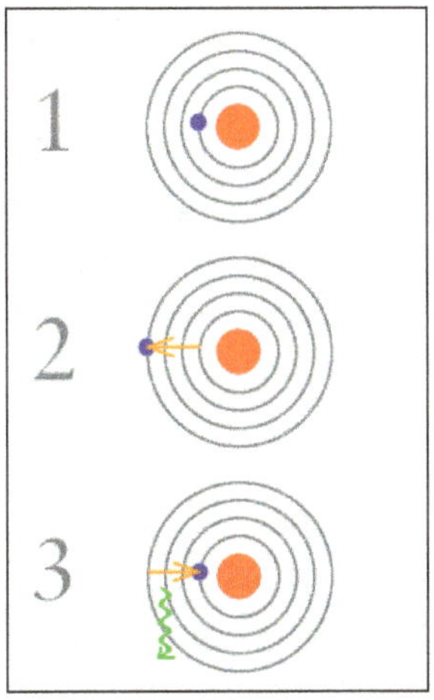

Schematics illustration of light emission from an atom.

Figure shows schematics of the mechanism behind light emission from an atom. The steps involved can be summarized as follows:

1. The atom is in its ground state with no electron present in excited states.

2. Light of a specific wavelength is incident onto the atom, which is absorbed by the atom, thereby promoting an electron from lower energy level to a higher energy level.

3. The electron returns to its original energy level (where it was present before excitation) by emitting a light photon. The color of the emitted photon or light depends upon the energy difference between the two energy levels involved in the process. As different atoms have different arrangements of their energy levels, dissimilar atoms emit light of varying colors.

A quantum dot also emits light by similar mechanism, because the charge carriers (electrons and hole) entrapped within it have discrete energy levels. However, the arrangement of energy levels within a quantum dot is also influenced by the dimensions of the dot. Thus a quantum dot made from same material can be tuned to emit light of different wavelengths or colors, depending upon the size of the dot. In this regard, bigger quantum dots emit light of longer wavelengths (or smaller frequencies), while smaller quantum dots produce light of shorter wavelength (higher frequency). These large dots give off red light and smaller dots produce blue light, and intermediate dots produce green light. In this way, by adjusting the size of the quantum dot, almost all the colors of the spectrum can be produced.

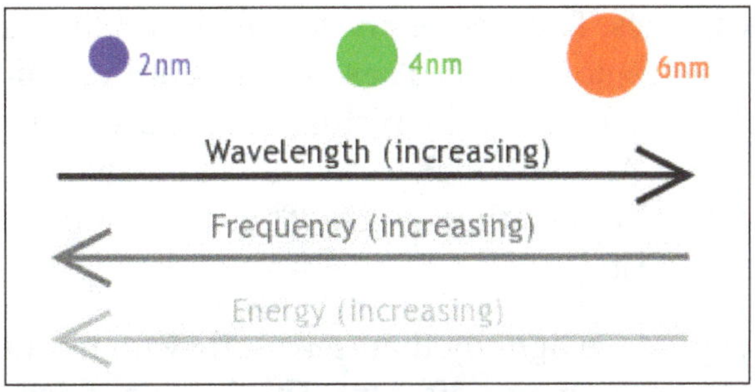

The size of the quantum dots is in nanometer range. Bigger dots emit light of longer wavelength and lower frequency, such that the color of the emitted light is close to red. Similarly, smaller dots produce shorter wavelength and higher frequency, and the colour is towards blue.

The band gap of a smaller quantum dot is more than that of a bigger quantum dot. Band gap can be described as the minimum energy required by the atom to promote an electron from the valance band to the conduction band. In this respect, a smaller quantum dot requires more energy to excite the electron. Since the frequency of the emitted light is directly proportional to the energy, small dots produce light having higher frequencies (or shorter wavelength). Similarly, larger dots more closely spaced energy levels, thus they produce lower frequency (or higher wavelength) light.

Preparing Quantum Dots

Quantum dots are nothing but small sized crystal. Therefore, they can also be produced similar to other crystals. Most popular techniques to produce quantum dots are: molecular beam epitaxy (MBE, wherein a beam of atoms is bombarded onto a substrate and the single crystal gradually builds up), ion implantation (wherein accelerated ion beam is incident onto the substrate), and X-ray lithography (atomic scale engraving using X-rays to produce the desired structures). Recently, several researchers have proposed preparing quantum dots by using biological processes, such as, by feeding metals to the enzymes.

Image showing a sensor chip of a webcam which converts the real time images into digital format when light is incident on its grid. The grid contains light detecting pixels. Quantum dots can be used to produce small sized pixels as well as sensors which have higher resolutions.

Quantum dots can also be used in computer screens and other displays, owing to the following important advantages:

- Conventional LCD (liquid crystal display) screens use a combination of red, blue and green crystals to show the images. These color crystals are basically color filters which can switch ON or OFF under electronic control. These crystals are illuminated from behind a very bright backlight. Thus the color of the picture is produced by combining tiny crystals of red, blur or green colors. As mentioned earlier, an individual quantum dot can be tuned to produce different colors by adjusting its size. Thus the colors produced by a quantum dot are likely to be more realistic.
- Since quantum dots produce light by themselves, they do not need any backlight to show images. Thus, they are more energy efficient. This is their added advantage which can benefit the portable devices which we frequently see around us, since they require less energy to perform.

- As the size of quantum dots is very small than that of the liquid crystal used in conventional LCD displays, the use of quantum dot is expected to produce much higher resolution image than possible by the LCDs.
- Quantum dots are much brighter than the recently used organic LEDs or the OLEDs. Thus, they can also be used to produce OLED displays.

Applications of Quantum Dots (QDs)

Solar Cells

One of the most important features of silicon of use in electronic devices (e.g. transistors) is doping. Wherein small quantities of various elements can be added to silicon in order to generate either an excess (n-type) or deficiency (p-type) of electrons in it, thereby enhancing the material's conductivity. Additionally, p-n junction can be created by joining n-type and p-type silicon, which are the building blocks for almost all the devices in electronic industry.

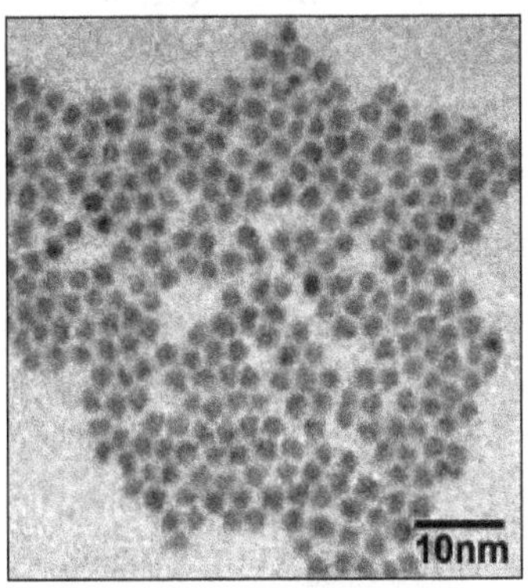

TEM images of InAs n-type QDs doped with silver, 3.3nm in diameter.

Doping the QDs is a relatively new area of research, and both n-type and p-type QDs have been produced. Figure shows the indium arsenide QDs which have been doped with silver to create n-type QD. Low cost solar cells can be realised by using QDs, as QDs can be easily prepared via simple and economic chemical processes. Additionally, thin-film photovoltaics can be prepared from QDs which have efficiencies comparable to those of traditional silicon cells. The enhancements in efficiency of QDs can be attributed to the different materials that can be used in their synthesis. Since some semiconductors can emit multiple electrons on absorbing one photon. In addition to this, the manipulation in their size and shape can result in absorption of light having different colours.

However, the synthesis of efficient QD based solar cells has not yet been possible. For preparing solar cell, an n-type, and a p-type nanocrystals are required. When light is incident on a solar cell, electrons and holes are generated as light photons get absorbed in the material. These electrons and holes must be separated so as to avoid their spontaneous recombination. The separation of these charge carriers results in the flow of electrons out of the semiconductor to the external electrical circuit. However, some of the electrons and holes recombine, this does not cause and electron (and thereby current) flow in the external circuit. This recombination is much pronounced in QDs than the large silicon crystals. Doping of the semiconductor nanocrystals for making p-n junction can separate the electrons and holes more efficiently. Conventionally, silicon is doped with P or B atoms, however, this cannot be done with the QDs because of their nanometer size. For comparison, a 4nm QD comprises ~1000 atoms. Addition of few dopant atoms can result in expelling these atoms from the nanocrystal.

Solar Cell Basics

In a typical solar cell, the absorption of light results in the generation of electron-hole (e-h) pairs. When the e-h pair is bound, it is termed as exciton. An internal electric filed (present in p-n junctions or Schottky diodes) can separate this pair, leading to the flow of electrons and holes, which in turn cause an electric current. This internal field can be generated by differential doping of one part of the semiconductor with electron donating atoms (n-type doping), and doping the other part with electron accepting atoms (p-type doping). This results in the formation of a p-n junction within the semiconductor. The e-h pair is generated only when the energy of the absorbing photon in greater than the bandgap of the material. Thus, (a) the photons of lower energy than the bandgap are not absorbed, and (b) photons with higher energy than the bandgap are quickly (within ~10–13 seconds) thermalised to the band edges, decreasing output. Part (a) causes a reduction in current, and thermalisation decreases the voltage. Consequently, a trade-off exists between voltage and current in a semiconductor cell. This can be partially avoided by employing multiple junctions. Complete balance evaluations envisage a maximum efficiency of 31% for a monolithic solar cell.

A little bit of mathematics can show that the maximum efficiency (i.e., 31%) corresponds to a bandgap of 1.3-1.4 eV, or the near infrared (NIR) spectrum. Silicon's bandgap (1.1 eV) is closest to this value, and this is one of the key factor in the predominance of silicon in solar cell applications. Notwithstanding this advantage, the maximum efficiency achievable with silicon is ~29%. This value can be enhanced via 'tandem' or, multi-junction approach, wherein multiple single-junction cells of different bandgaps are vertically stacked. As an example, a 2-layer cell should have one layer of 1.64 eV and second with 0.94 eV, resulting in maximum efficiency of 44%. Similarly, a 3-layered (1.83, 1.16, 0.71 eV) gives the value of 48%. Theoretical calculations have predicted the maximum efficiency of ~86% for an infinite layered cell, where the thermodynamic losses limit the efficiency to this value.

What about Quantum Dots

QDs are semiconductor particles whose dimensions are less than the exciton Bohr radius, resulting in quantum confinement effect being dominant in deciding their properties. Due to the confinement effects, the electron energy levels become finite. The situation is similar to an atom, and that is why, QDs are often termed as 'artificial atoms'. By adjusting the size and shape of the QDs, these energy levels can be tuned, which then defines the bandgap. QDs can be made with different sizes, thereby changing their bandgap without any modification in the underlying material or the synthesis processes. Ideally, the size of the dot can be changed by simply controlling the synthesis time and temperature.

Owing to the tunable bandgap, QDs are interesting for solar cell applications. Single junction cells of PbS (lead sulfide) colloidal QDs demonstrate bandgaps in the range of far infrared, which is difficult to obtain in traditional cells. Since the maximum portion of the solar energy reaching the Earth is in infrared and near infrared regions, a QD, owing to its bandgap in the right spectrum can make use of this enormous energy. Furthermore, colloidal QDs can be easily synthesised. Being suspended in a colloidal liquid, they are easily handled throughout production process. Colloidal QDs can be produced in large amounts and the dots can be spread on desired substrate via spin coating.

Dye Sensitized Solar Cell (DSSC)

These are the most recent cell design, wherein a sponge like layer of TiO_2 is used as semiconductor value and also provides mechanical support. During construction, an organic dye (e.g., ruthenium-polypyridine) is filled in the sponge which injects electrons into TiO_2 upon photoexcitation. Since ruthenium is a rare metal, the cost of the dye is high. Since the discovery of DSSC, the use of QDs as a substitute to the organic dyes has been considered. Owing to the possibility of tuning the bandgap, other design parameters (such as material sued for other parts of the cell) can also be changed. For instance, the researchers have developed a cell wherein the rear electrode is in contact with a film of QDs, thereby eliminating electrolyte, and resulting in a depleted heterojunction. The cell had an efficiency of 7.0%, which is superior to solid-state DSSC devices, but less than devices using liquid electrolytes.

Multi-junction

In cells absorbing multiple frequencies, cadmium telluride (CdTe) is usually used. A colloidal suspension of CdTe crystals in spin coated over a substrate (e.g., glass slide). Such cells do not employ QDs, since low scale production of QDs is expensive.

Hot-Carrier Capture

The efficiency of the cell can also be improved by capturing the surplus energy of the electron when emitted from a single bandgap material. In materials like silicon, the emission site is at a fair distance from the electrodes (where electrons are collected); electrons interact with the crystal lattice, dissipating their surplus energy as heat. So the electrons do not have extra energy. For increasing these interactions, amorphous silicon was studied; however, the large amount of defects inherent to amorphous materials suppressed their advantages. Modern thin-film cells have less efficiency than conventional silicon cells. To alleviate these problems due to defects, thin uniform films can be prepared from nanostructured donating species. However, these structures suffer from other issues inherent to QDs, e.g., high resistivity, heat retention.

Multiple Excitons

Researchers in 2004 reported that in QDs, multiple excitons can be created by absorption of one photon. Thus, more solar energy can be captured by collecting them. This approach is termed 'carrier multiplication' (CM) or 'multiple exciton generation' (MEG), wherein the QD is modified to generate more than one e-h pairs at relatively lower energy than one pair at high energy. This results in enhanced photocurrent, thereby increasing the efficiency.

Non Oxidizing

Recently, researchers from MIT developed solar cells from ZnO/PbS which are stable in air and have an efficiency of 9.2% (in laboratory conditions). The high efficiency was the result of efficient light absorption and charge transport to the electrodes (current collector). The cells are highly stable in air and retained their performance for more than 150 days of storage in air.

Display Devices

Quantum dot display employs semiconducting nanocrystals or QDs as display element, where QDs either emit or convert (in case of LED backlit LCDs) light. Presently available display devices (e.g., TV), termed as QLED, use QDs to convert light for LCD backlights, rather than using them for actual display. The use of QDs as light source was first proposed in 1990s, and typical applications were imaging with QD infrared photodetectors, LEDs, etc. Presently, QDs are intensely investigated for their use in light sources and displays. By controlling the composition and size of a quantum dot, its properties can be controlled. Furthermore, quantum dots are photoactive (photoluminescent) as well as electroactive (electroluminescent), thus, they can be easily incorporated in new emissive display devices. True QD displays (QLEDs) are at experimental stage and are different from the present commercial devices. QLEDs use QDs or semiconductor nanocrystals as electro-optical display technology, similar to organic LEDs or OLED displays where light is emitted when required. Thus these displays are efficient and the next technology after OLEDs. There are concerns over the use of cadmium to produce QLEDs, thus such devices are still experimental. Big and flexible displays can be prepared by using QDs. Besides, these displays do not easily degrade like the OLEDs, thus, quantum dots are potential candidates for flat-panel TV screens, digital camera, mobile phones, etc.

Optical Properties of QDs

In contrast to atoms, the energy levels in a QD exhibit strong dependence upon its size. For instance, CdSe QDs emit light of different colours for different sized dots, i.e., a 5nm dot produced red, and 1.5nm dot produced violet light. These changes in colour depending on the size can be attributed to the quantum confinement effects and directly depends on the energy levels of the dot. The energy (and color, in turn) is determined by the bandgap of the QD which varies inversely with the square of the QD size. Thus, larger dots have more energy levels and these are closely spaced, these QDs emit or absorb lower energy photons (redder color). In brief, the energy of the emitted photons increase with decrease in the size of the QD, since more energy is needed to confine the excitation to a smaller volume.

Quantum Dot Light Emitting Diodes (QLEDs)

QLEDs emit characteristically pure and saturated colors having narrow bandwidths. The wavelength of the emitted light can be easily controlled by adjusting the QD's size. The other advantages of QLEDs include high efficiency, flexibility, relatively lower processing costs than OLEDs. QLEDs can be tuned to operate in the complete visible range, i.e., from 460nm (blue) to 650nm (red). Additionally, by altering the chemical composition of the quantum dots, the emission wavelength can further be extended to UV as well as NIR wavelengths. QLED technology involves using electroluminescent nanoparticles instead of photoluminescent QDs used in present quantum dots based TV. Thus, in QLED, direct emission of light is used as display, rather than conversion via LED backlights. Instead of using a separate backlight for illumination, QLED TV locally controls the light emitted from individual pixels.

Nanoelectronics

Nanoelectronics involves using nanotechnology in electronic components. The devices produced are so small that interatomic interactions and quantum confinement effects are applied to them. The devices include hybrid molecular/semiconductor electronics, nanotubes/nanowires (e.g. SiNWs, CNTs), etc. Recent silicon CMOS devices are also within this length scale. Nanoelectronics is often termed as disruptive technology as present candidates are significantly different from traditional transistors.

Basic Concepts of Nanoelectronics

Gordon Moore, in 1965, proposed a continuous down scaling of silicon transistors, this was later known as Moore's law. Since then the transistor feature size has been reduced from 10 microns to around 25nm till 2011. Nanoelectronics is bout continuously realising this law via novel materials and techniques to fabricate electronic devices of nanoscaled dimensions. An important mathematical relation highly important in nanoelectronics states that, 'an object's volume decreases as third power of its linear dimensions, while its surface decreases as the second power'. Thus, the downsizing of objects in nanotechnology must be assessed properly, taking into consideration the above mentioned rule.

Nanoelectronic Devices

Present manufacturing processes use conventional top-down approach for device fabrication. Using this technique, critical length scales in ICs has already reached the nanoscale, for instance, the gate length in transistors widely used in modern electronic devices is of the order of 50nm.

Memory Storage

Traditional memory devices use transistors for storing information. With the help of cross bar switches based electronics, ultra high density storage devices can be produced which have reconfigurable interconnections between vertical and horizontal wiring arrays. This has been achieved by Nantero (developed CNTs based crossbar memory 'Nano-RAM'), and Hewlett-Packard (proposed memristor to replace flash memory).

Novel Optoelectronic Devices

Optical and optoelectronic devices offer extremely large bandwidths and high capacities; and are replacing the conventional analog devices in modern communications systems. These include photonics crystals and QDs. In photonic crystals, the refractive index varies periodically with a lattice constant. The lattice constant is the half of the wavelength of light used. They behave like a semiconductor, with the exception that they work with light or photons, rather than the electrons. Thus, photonic crystals have a tunable bandgap for propagating a specific wavelength of light. QDs are nanosized objects which can also be used to construct lasers. A QD based laser offers the advantage of tunable emission wavelength over the conventional semiconductor lasers. The emission wavelength can be manipulated by changing the diameter of the QD. Additionally, QD lasers are inexpensive and provide high beam quality.

Quantum Computers

Quantum mechanical principles can be exploited in quantum computers, thereby enabling the use of fast quantum algorithms. Quantum computers use quantum bit (qubit) as memory space for simultaneous multiple computations. Thus much faster computers can be build.

Energy Production

Nanowires and other nanostructured materials can be used to produce inexpensive as well as highly efficient solar cells. More efficient use of solar energy is imperative towards meeting the global energy requirements. Another application of nanostructured materials in energy production is the use of bio-nano generators which can operate in vivo. These are nanoscaled electrochemical devices analogous to fuel cells. They draw power from the glucose present in the blood of living body, in a similar manner as the body uses food to generate energy. This is achieved by using an enzyme which removes the electrons from the glucose, these free electrons are then used in the device. A bio-nano generator can be used to generate ~100 Watts power (~2000 food calories per day). This is true if the complete food is converted into electricity. Since human body also need some energy forits essential functions, actual electricity produced via bio-nano generator could be less. However, this electricity can be used to power devices embedded within the body (e.g., pacemaker).

Medical Diagnostics

Nanoelectronic devices for the detection of the concentration of biomolecules in realtime presents an attractive application as medical diagnostics. Similarly, devices can be produced which can interact with individual cells. Such devices are known as biosensors and represent a recent area of research for nanoelectronic devices. These applications of nanoelectronic devices are important for health monitoring, defense technology, etc.

Nanowire

Nanowires are based up on a flat substrate of semiconductor materials, such as silicon and germanium. Nanowires are simply very tiny wires. They are composed of metals such as silver, gold or iron. Nanometer is measured as spatial measurement that is about 10^{-9} meters which are mostly used in nanotechnologies for the manufacturing of nano machines. The small nanowire is created by Nano particles with a diameter as small as nanometer. Basically the diameter of nanowires is one nanometer, engineer's works with 30 and 60 nanometers.

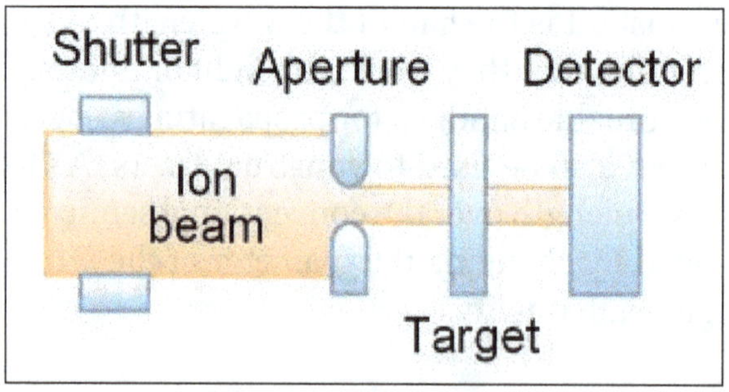

Figure is consisting of ion beam rod on which shutter, aperture, target and detector are mounted on it which is in the form of tube. A nanowire plays a significant role in field of quantum computers and nanorobots are very tiny machines which are planned for a specific function or tasks repeatedly with some accuracy at nonascale dimension. A wide variety of elemental, binary and compound semiconductor nanowires has been synthesized via the VLS method, and comparatively good control over the nanowires diameter and diameter distribution has been achieved.

There are two basic approaches to synthesizing nanowires: top-down and bottom-up. A top-down come within reach of reduce a large piece of substance to small pieces. A bottom-up approach synthesizes the nanowire by combining component ad atoms. Most synthesis techniques use a bottom-up approach. Nanowire transistors made with conventional lithographic fabrication methods can improve performance in nonascale electronics. There are different types of nanowires in technology, they are: Metallic nanowires, semiconducting nanowires, insulating nanowires. The structure of nanowires is very simple, made of variety of materials.

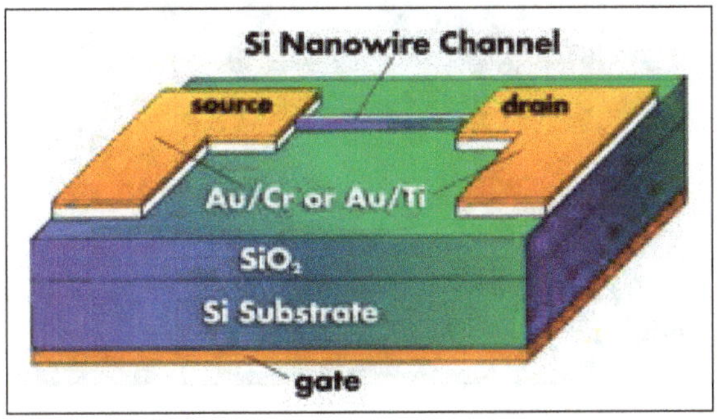

A simple Silicon Nanowire Transistors is shown in figure. Silicon nanowire transistor simplifies both processing and allows the devices to be switched on and off more easily. The 60 nanometer wide channels exhibit a much greater difference in current between the on and off states than is true for larger reference channels up to 5 micrometers wide. This suggests that when a channel is scaled down to the nano regime, the ultra-narrow proportions significantly reduce the current leakage associated with defects in silicon. As a result, the transistors are less sensitive to electronic noise in the channel and can be turned on and off more effectively.

Properties of Nanowire

- Mechanical property: The enormous amount of grain boundaries in a bulk material are made of nanoparticles which allow extending the grain boundaries sliding leads to high flexibility. The figure below consists of gate insulator device and substrate which involve in the operation of the mechanical property of nanowire.

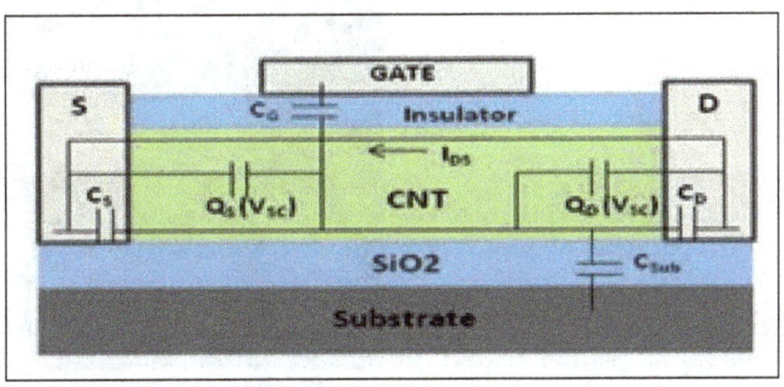

- Magnetic property: In the magnetic property of nanoparticles the energy of magnetic anisotropy might be that miniature that the vector of magnetization fluctuates thermally, this is called super magnetism. Such materials are free from reminisce and coercitivity. Touching super magnetic particles are loosing this special property by get in touch with expect the particles are kept at distance. Unusually electronic and magnetic characteristics are established at non zero temperature such as the metal insulator alteration in metal oxides non Fermi liquid performance of highly interrelated f-electron compound, uncharacteristic symmetry state of high-Tc superconductor device. Combining the particles with high energy of anisotropy with a super magnetic may over leads to a new class of permanent magnetic materials.

- Catalytic property: Due to large surface of area, the nanoparticles which are made of transition materials oxide exhibits motivating catalytic properties. In some of the special cases catalysis may be improved and additional specific by decorating these particles with gold and platinum duster.

- Optical property: In optical property the allotment of non-agglomerated nano particles in a polymer are used to the directory of refraction. In addition, such a procedure may manufacture material with non linear optical properties or visual property. Gold and Cd se nano particles in glass leads to red or orange coloration semi conducting nano practices and some oxide polymer nano compositor exhibits fluorescence performance is blue shift with decreasing particles size. Faraday rotation is one of the magneto optical effects extremely premeditated for Ferro fluid.

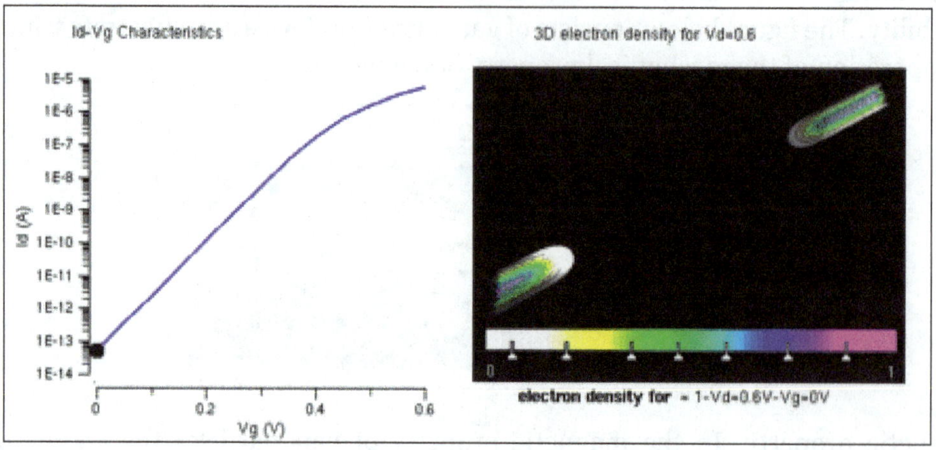

Applications of Nanowires

Nanowire devices can be assembled in a rational and predictable because:

- Nanowires can be precisely controlled during synthesis.
- Chemical composition.
- Diameter.
- Length.
- Nanowires are used in hertostructure which are sub divided as axial hetrostructure for e.g.- Gap-GaAs, radial hetrostructure e.g.- SiGe and nano superlatticles.
- Nanowires are mostly applicable in sensors like ph sensor and Gas sensor.

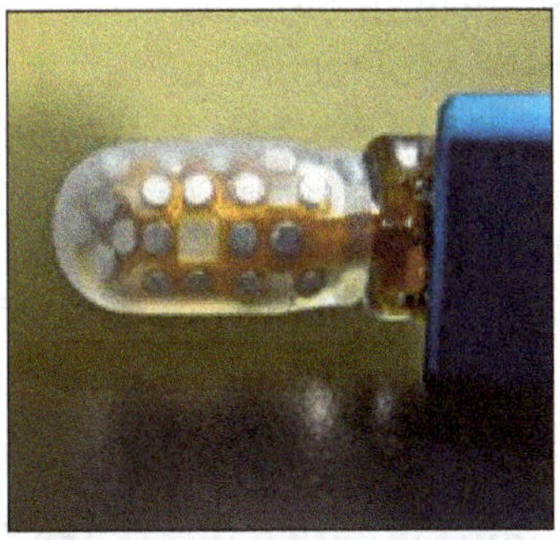

- Used in the manufacturing of nanophotons and nanoprobes with high temperature and high laser technology.
- Reliable methods exist for their parallel assembly.

Nanowires are represents the best-defined class of nanoscale building blocks and this precise control over key variables has correspondingly enabled a wide range of devices and integration strategies to be pursued.

Semiconductor and Oxide Nanowires

Semiconducting nanowires are of great interest due to their potential in electronics, optoelectronics, and sensor applications. Silicon nanowires (SiNWs) have received the most attention due to the historical role of silicon in the integrated circuits (IC) industry. Continued high performance from silicon may require alternatives to thin film based planar CMOS transistors in the form of novel devices, three-dimensional structures, innovative architectures, and integration with other functional components. SiNWs may provide new avenues in these directions. Silicon becomes a direct band gap semiconductor at nanoscale dimensions and thus may have interesting optoelectronics applications. The interest in optical interconnects beyond the copper era in IC manufacturing and the desire to have all silicon-based process (instead of using III–V materials for the optical interconnect part) may benefit from mature nanowire technology if and when it becomes available. In many of the applications, Ge provides a competitive avenue and thus received serious consideration in the nanowire community. Nanowires of III–V semiconductors such as GaAS,

AlxGa1−xAs, InP, InAs, and wide band gap III−V nitrides have also been reported. a wide variety of techniques including the VLS approach, laser ablation, template-guided synthesis, oxide-assisted growth, and others have been used to grow semiconducting nanowires.

Silicon Nanowires

Silicon nanowires, also referred as SiNWs, are a type of nanowire most often formed from a silicon precursor by etching of a solid or through catalyzed growth from a vapour or liquid phase. Initial synthesis is accompanied by thermal oxidation steps to yield structures of accurately tailored size and morphology. SiNWs shows many unique properties that are missing in bulk (3D) silicon materials. These properties arise from an unusual quasi 1-D electronic structure and are the subject of research across numerous disciplines and applications. The reason that SiNWs are considered one of the most important 1D materials is that they could have a function as building blocks for nanoscale electronics assembled without the need for complex and costly fabrication facilities. SiNWs are frequently studied towards applications including photovoltaics, nanowire batteries, thermoelectrics and non-volatile memory. A large number of techniques exist to fabricate silicon nanowires. These can be classified into bottom-up and top-down fabrication techniques. In top-down fabrication, lithography is used to define the fabricated structure that is then transferred from the photo-resist to the substrate by etching or a similar way of structuring the already available material. In the bottom-up approach, the material is added to the substrate in a self-organized way.

Germanium Nanowires

Germanium is an important semiconductor with a direct band gap of 0.8 eV and an indirect band gap of 0.66 eV. It has been receiving much attention recently either as a pure element or as an alloy with silicon ($SixGe_{1-x}$) for logic and other devices. Germanium offers higher intrinsic carrier mobilities than silicon:

$$\left(\mu_n = 3900,\ \mu_p = 1900\ \text{cm}^2/\text{V s}\right) \text{versus} \left(\mu_n = 1500,\ \mu_p = 450\ \text{cm}^2/\text{V s}\right)$$

At room temperature, enabling faster switching and higher frequency devices. Other advantages include higher intrinsic carrier concentration, 2.4×10^{13} cm^{-3} versus 1.45×10^{10} cm^{-3}; larger bulk excitonic Bohr radius, 24.3 nm versus 4.7 nm for more prominent quantum confnement effects even at larger material dimensions; and compatibility with high-k dielectrics enabling integration to current semiconductor processing technology. Germanium is also compatible with III–V materials, for example, good lattice matching with GaAs. Germanium oxide exhibits interesting optical properties suitable for interesting integrated optoelectronic circuits. All these features have prompted extensive investigations of germanium nanowire (GeNW) growth, characterization, and application development.

III-V Nanowires

III–V compound semiconductor materials are of great interest in electronics and optoelectronics for applications in high-power or high-frequency transistors, light emitting diodes, lasers, resonant tunneling diodes, solar cells, photodetectors, and others. Their usefulness in the above applications has been well established in the last two decades through thin flms and quantum wells of GaAs, $Al_xGa_{1-x}As$, $In_xGa_{1-x}As$, InP, GaP etc. grown using techniques such as metal organic chemical vapor deposition (MOCVD) and molecular beam epitaxy (MBE). Recently, these materials have also been grown as onedimensional nanowires since quantum confined low-dimensional structures have the potential to provide superior performance in all of the above applications.

- GaAs Nanowires: Gallium arsenide is a direct band gap semiconductor with a band gap of 1.424 eV, electron saturation velocity of 10^7 cm/s, and carrier mobility of 8500 and 450 cm^2/V s for electrons and holes, respectively. It has been grown in the nanowire form by laser-assisted catalytic growth, MBE, MOCVD and template-guided approach.
- InAs Nanowires: InAs exhibits a small energy band gap (0.35 eV), high electron mobility (33000 cm^2/V s), and a large quantum confnement energy. There have been several reports on the preparation of InAs nanowires. InAs growth by MOCVD with a thin layer of SiO serving as catalyst on InP(111), InAs(001), and Si(001) substrates. Figure shows SEM images of InAs nanowires on InP substrates at temperatures 540 °C–660 °C.

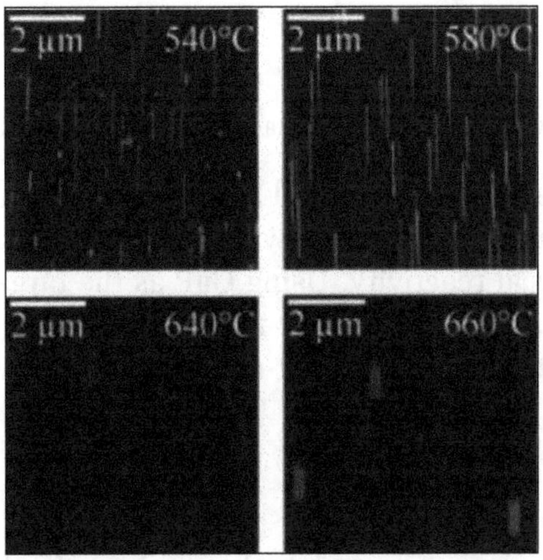

InAs nanowires grown on InP(111)B by MOCVD at various temperatures. Growth conditions: pressure = 10 kPa, TMIn mole fraction = 2 × 10^{-6}, AsH_3 mole fraction = 2 × 10^{-4}, H_2 carrier gas glow = 6 L/min.

- InP Nanowires: Indium phosphide is a direct band gap semiconductor with a gap of 1.34 eV that has been widely investigated and used for electronics and optoelectronics applications. It has been grown in the form of nanowires using laser-assisted growth, sublimation of InP powder, MBE, and MOCVD. Figure shows results of InP nanowires synthesized using MOCVD technique, which provides vertical, uniform diameter nanowires.

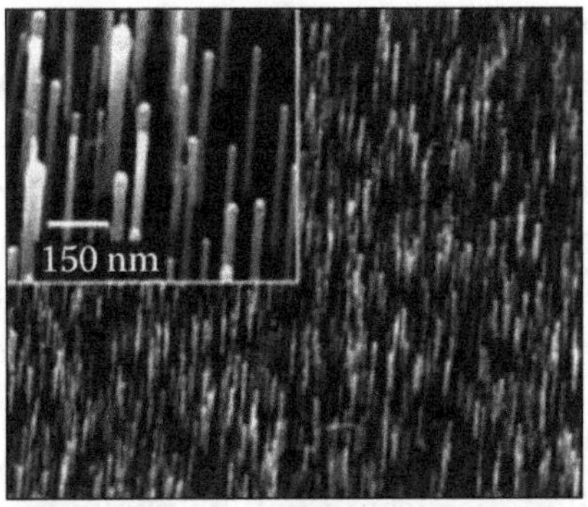

InP nanowires grown by MOCVD using 20 nm gold colloids.
TBP/TMIn ratio = 120 and 430°C growth temperature.

- GaP Nanowires: Gallium phosphide is a wide band gap semiconductor with an indirect gap of 2.26 eV. The electron mobility of bulk GaP is 160 cm^2 /V.s. GaP is preferred for high-temperature electronics applications and light emission in the visible range. In laser catalytic growth of GaP, a GaP solid target is laser ablated to provide the source vapor and the use of gold colloids appears to provide a tight diameter distribution of grown nanowires. In general, the mean nanowire diameter is only 1–2 nm larger than that of the colloids. Sublimation of GaP powder has also been used to generate the vapor source.

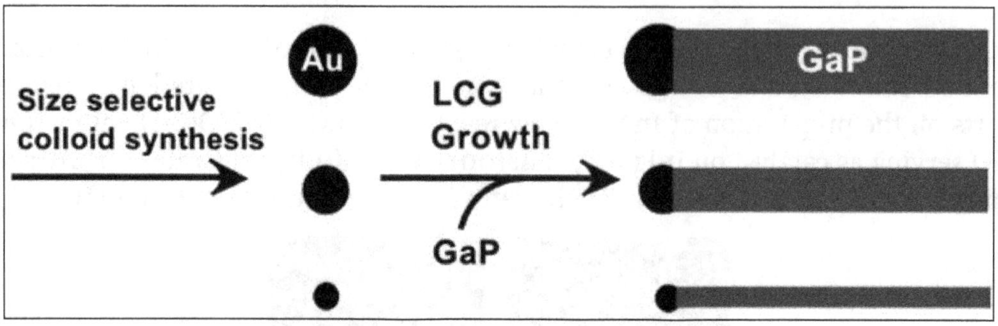

Schematic of laser ablation grown well-defined GaP nanowires with monodisperse gold colloids as catalysts.

Laser ablation is a laser-assisted method and the key feature of this technique is that the catalyst used to define nanowire VLS growth can be selected from phase diagram data and/or knowledge of chemical reactivity. Using GaP as the target, GaP nanowires on well-defined gold colloids coated substrates by the laser ablation growth was grown. Figure is schematic of laser ablation growth of well-defined GaP nanowires using monodisperse gold colloids as catalysts.

Oxide Nanowires

Metal oxides are an important class of materials that find applications in a wide variety of fields. For example, ZnO is a direct, wide band gap semiconductor with applications in ultraviolet (UV) and near UV photonics. Many metal oxides find applications either as catalyst supports or active catalyst materials and conducting channels in gas sensors. Several metal oxides also find direct applications in electrochemical energy devices such as lithium ion batteries, electrode materials, and solar cells. Almost all of the above applications can benefit from the availability of metal oxides in the form of nanowires due to various reasons: well-defined crystallinity and surface, reactivity, electronic properties, and high surface area. In some cases, metal oxide nanowires with diameters less than few nanometers are sought for understanding the quantum confinement effects. Even though, nanowires with radius larger than the corresponding Bohr radius may not show distinct properties compared to bulk, they are still promising due to the following reasons: fast charge transport, low thresholds for percolation, high surface area, and well-defined crystal surfaces. The above-mentioned applications are diverse and require metal oxide nanowires in different formats ranging from epitaxial nanowire arrays, vertical nanowire arrays on a variety of conducting substrates, networked thin films, highly dense thick flms, and powders. For epitaxial nanowire arrays, there should be a minimal lattice mismatch between the nucleating metal oxide crystal and the underlying single-crystal substrates. For vertical nanowire arrays, high nucleation densities are typically necessary. A more detailed discussion will be provided under different categories of the metal oxide nanowire growth schemes.

Synthesis Methodologies

In the case of oxide nanowires, there are four important vapor-phase methodologies used for their synthesis:

- Foreign metal catalyst assisted.
- Direct oxidation of large, molten metal droplets.
- Direct chemical vapor transport or chemical vapor deposition (CVD).
- Thermal/ plasma oxidation of metal foils.

Catalyst Assisted

The CVD of metal oxides when performed on substrates coated with a thin film or layer of catalyst particles can lead to metal oxide nanowire growth with catalyst particles at their tips. However, the metal containing gas or liquid precursors are not readily available for deposition of many metal oxides. So, it is easier and inexpensive to start with either solid metal or metal oxide powder sources. The catalyst is typically deposited onto substrates as thin films (<100 nm thickness) either using sputtering or evaporation.

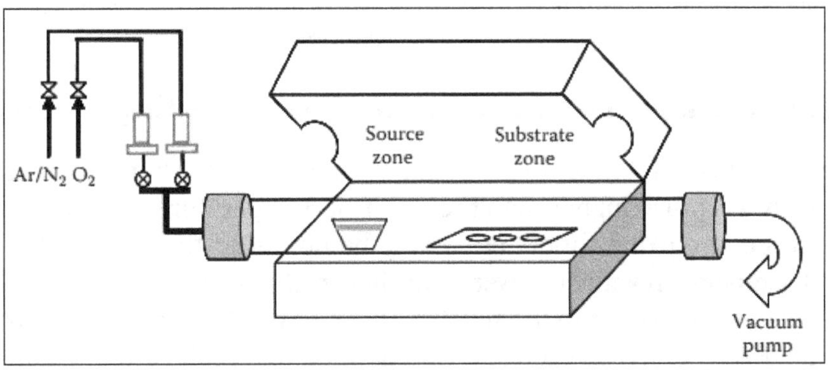

A schematic showing a typical horizontal quartz tube reactor used for the catalyst-assisted synthesis of nanowires.

In some cases, dispersions of catalyst particles can be synthesized or purchased separately and spin-coated onto the substrates. Many of the catalyst-assisted synthesis techniques for metal oxide nanowires use thermal CVD in a quartz tube reactor as shown in the schematic in Fig. This setup includes a horizontal tube heated with a two-zone furnace with one zone for evaporation of solid metal sources and another zone for growth on substrates coated with the catalyst layer.

Direct Oxidation Schemes using Low-Melting Metals

The synthesis procedures involving direct oxidation of metals are gaining importance compared to catalyst-assisted schemes mainly because of two reasons: possible contamination with foreign catalyst metals in the catalyst-assisted schemes and the costs associated with the respective catalyst metals and corresponding processes. The solubilities of oxygen and the respective oxides in the low-melting point metal melts are expected to be negligible. So, any amount of dissolution of oxygen species during direct oxidation of large, micron-scale molten metal droplets leads to spontaneous nucleation of nanometer scale, metal oxide nuclei from molten metal. Further growth of metal oxide nuclei via basal attachment leads them to one-dimensional structures. High density of nucleation followed by growth via basal attachment (from dissolved oxygen) from large, molten metal clusters results in flowery morphology in which a high density of nanowires emanates from one point. After nucleation stage, further growth of metal oxide nuclei can also occur laterally in addition to basal growth. Such growth behavior leads to polycrystalline crust formation on large molten metal clusters. It has been found that the presence of reducing species such as hydrogen or chlorine reduces the lateral propagation of metal oxide nuclei. This is schematically illustrated in figure.

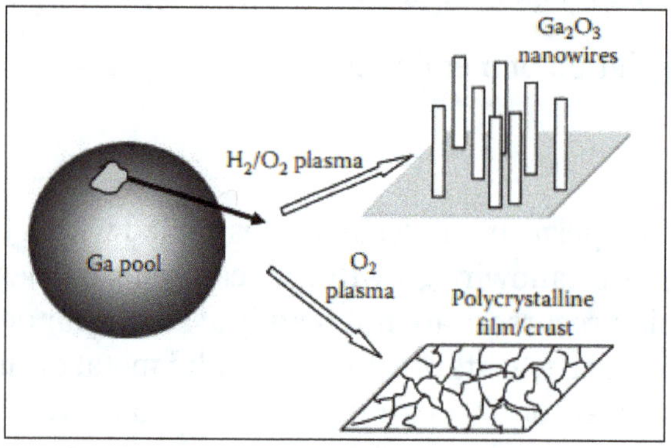

A schematic illustrating the effect of hydrogen or other reducing gas-phase precursors during direct oxidation schemes for bulk nucleation and growth of oxide nanowires from their respective molten metal droplets using Ga as an example.

Chemical Vapor Transport or Deposition of High-Melting Metal Oxides

In the case of high-melting point metals (for example, tungsten, tantalum, and molybdenum), their melting points are always much higher than the synthesis temperatures used in any procedure. For such metals, the respective oxide vapors can easily be produced through oxidation of the corresponding metal at temperatures much lower than their melting points of metals. The transport of the respective metal oxide vapors onto the substrates can be used to create metal oxide films similar to any CVD scheme. However, it has been shown that under particular circumstances with low oxygen partial pressure, such chemical vapor transport procedures, resulted in the synthesis of respective oxide nanowires without the aid of any external catalyst. Such a scheme can be appropriately described as "vapor–solid" since the respective metal is not molten at the synthesis temperature.

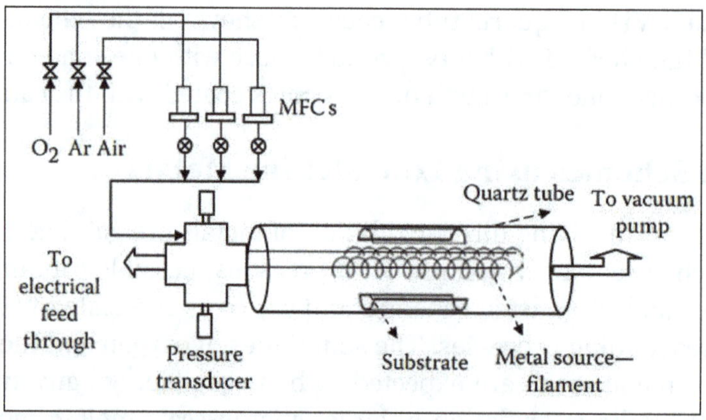

A schematic of the hot-wire CVD reactor setup used for chemical vapor transport experiments using a variety of high-melting point metals as filaments.

The hot wire (or filament) CVD experiments were conducted using oxygen flow over tungsten filaments for the synthesis of tungsten oxide nanowires. As shown in figure, the reactor consists of a 2 in. diameter quartz tube with the metal f laments wound on top of two ceramic tubes, which are connected to an electrical feed through. The filament is resistively heated by flowing electric current and the temperature is controlled by controlling the applied power to the filament using a variac power supply. The quartz tube is placed inside a furnace oven. In addition to the oven, the radiation from the filaments can heat the substrates to high temperatures depending upon their distance from the source. The substrates can be easily heated up to temperatures around 550°C in 2 in. quartz tube without the use of external heating via furnace. Experiments using tungsten filaments and oxygen flow at low partial pressures in the range of 0.05 to 0.3 Torr, filament temperature around 1600°C and substrate temperatures around 400°C to 800°C resulted in the synthesis of tungsten oxide nanowires.

Plasma and Thermal Oxidation of Foils

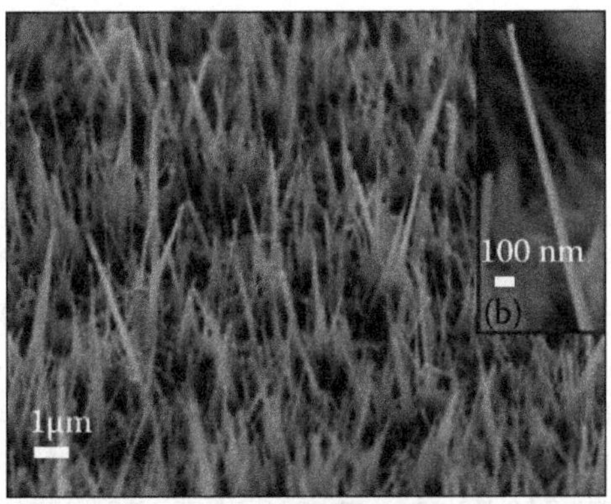

SEM images of iron oxide nanowires synthesized through plasma oxidation of an iron foil: (a) side view of an iron oxide nanowire array on top of the iron foil substrate; and (b) inset shows SEM image of a single tapered iron oxide nanowire.

Another concept that is interesting for synthesizing metal oxide nanowire arrays involves direct oxidation of metal foils. Here, the metal foils are oxidized either thermally or using highly dissociated oxygen plasmas. The underlying nucleation and growth modes in these cases are different from that of "tip-led" growth in catalyst-assisted and self-catalyzed schemes but are somewhat similar to that of bulk nucleation and growth from low-melting metal melts. The main difference is however that the nanowire growth occurs here with direct oxidation of solid metal foils and at temperatures lower than the melting temperatures of the respective metals. Plasma oxidation of iron foils using highly disassociated oxygen plasma was created using inductively coupled radiofrequency (RF) plasma. These oxidation experiments were conducted in a vacuum chamber equipped with an inductively coupled RF generator with a maximum power of 5 KW, reactor pressure of 2 Pa, and experimental duration of about 2 min. The resulting nanowire arrays are shown in figure.

Nanofiber

Nanofibers are incredibly small, synthetic fibers each measuring one thousand times thinner than a human hair. Nanofibers form complex strength through spiderweb-like structures, which have an advantage over conventional or even natural fibers due to their extraordinarily large surface area and tiny pore size. This combination creates maximum air flow and particle capture significantly more than other synthetic or natural filters, achieving greater efficiency down to PM0.3 microns. This image helps us visualize the size of a nanofiber when compared to both hair and a speck of pollen.

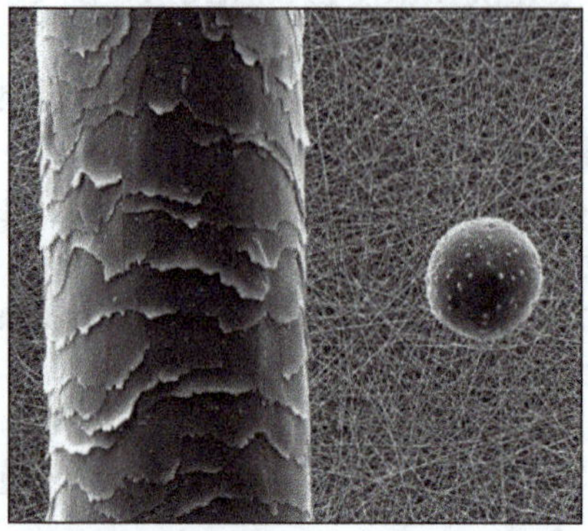

Its small size provides an array of benefits when utilized in face mask technology including advanced protection and better comfort. Featuring a weaving design closely packed together so microscopically that we are unable to see them with the human eye, nanofibers craft a small pore size and high surface-to-volume ratio. This means the surface area and pores of a face mask with nanofiber technology will not allow cells to pass through to reach the wearer. Simply, due to the unique and tough threading of the nanofibers, everyday dangers such as pollution, viruses, bacteria and more will be unable to cross the threshold of the mask, thereby enhancing protection from the outside world. Not even water or liquid can penetrate the surface, which is why it is such an essential component of any truly effective face mask. Additionally, this works two ways, as a mask is not only a protector of the individual wearing the mask, but it also serves to halt the spread of germs as well. When you're sick, you should have a mask on especially if there are other people in your household or if you are traveling out to an appointment with a doctor. Nanofibers block larger cells from escaping the mask, which helps to ensure less spread of any unpleasant or harmful ailment to unsuspecting passersby or loved ones.

While many are aware of the advanced protection that nanofibers provide in face mask technology, what may come as a surprise is that this material is also significantly more comfortable than other fibers. The complexity of the microscopic hair-like design makes the mask lightweight and, therefore, extremely breathable. It's not simply a cloth over your face it's a material that works with your body to trap cells out but allow air to flow in, so you don't feel cramped or suffocated. This also allows for better temperature regulation. More airflow means less of a "humid" feeling, as if the mask were a part of your face rather than your face feeling as if it's sweating into the mask. The light, soft and airy material of nanofibers ensures both short-term and long-term wear is much more comfortable. Overall, nanofibers have revolutionized the way we keep people safe with transformational applications in healthcare advancements such as tissue engineering, wound dressings, face masks and more.

Nanotube

A nanotube is a nanoscale material that has a tube-like structure. Nanotube structures have many applications in the general field of nanotechnology, which is a relatively recent field with much potential, as well as some significant liabilities. A nanotube can be made from various materials such as carbon, boron or silicon. Each of these has its own applications to new technologies. Carbon nanotubes, for instance, have been found useful as an agent for certain kinds of coatings for consumer products. In addition, carbon nanotubes (and silicon nanotubes) can be useful as a conductor in nanoscale projects or as an agent for measuring certain kinds of frequencies. Although nanotubes have a lot of potential, like other nanoscale projects, there are big questions about risks concerning coherence or cohesiveness, such as whether nanomaterials are safe and whether they may stay together or break off into tiny airborne bits. These are some of the issues about nanotubes and other nanomaterials as companies look at bringing new products and services to market that use nanotechnology.

References

- Quantum-Dot-An-Overview: inflibnet.ac.in, Retrieved 09, February 2020
- Nanowire: elprocus.com, Retrieved 17, June 2020

3

Nanoparticles and its Types

The particles of matter whose size lies between 1 and 100 nanometers are known as nanoparticles. Some of its prominent examples are magnetic nanoparticles, radioactive nanoparticle, nanosponges, nanofluid, magnetic-plasmonic bifunctional nanoparticles and nanoshell. The diverse aspects of these nanoparticles have been thoroughly discussed in this chapter.

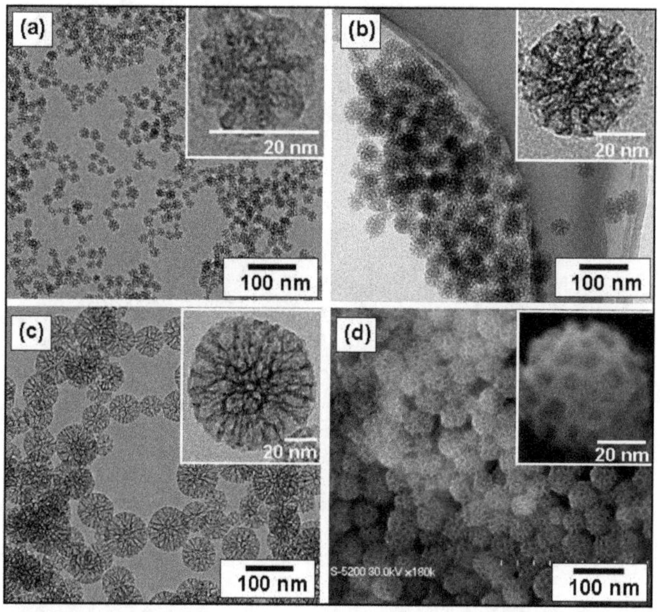

TEM (a, b, and c) images of prepared mesoporous silica nanoparticles with mean outer diameter: (a) 20nm, (b) 45nm, and (c) 80nm. SEM (d) image corresponding to (b). The insets are a high magnification of mesoporous silica particle.

Nanoparticles are particles between 1 and 100 nanometres (nm) in size. In nanotechnology, a particle is defined as a small object that behaves as a whole unit with respect to its transport and properties. Particles are further classified according to diameter. Ultrafine particles are the same as nanoparticles and between 1 and 100 nm in size, fine particles are sized between 100 and 2,500 nm, and coarse particles cover a range between 2,500 and 10,000 nm.

Scientific research on nanoparticles is intense as they have many potential applications in medicine, physics, optics, and electronics. The U.S. National Nanotechnology Initiative offers government funding focused on nanoparticle research.

Definition

The term "nanoparticle" is not usually applied to individual molecules; it usually refers to inorganic materials.

The reason for the synonymous definition of nanoparticles and ultrafine particles is that, during the 1970s and 80s, when the first thorough fundamental studies with "nanoparticles" were underway in the USA (by Granqvist and Buhrman) and Japan, (within an ERATO Project) they were called "ultrafine particles" (UFP). However, during the 1990s before the National Nanotechnology Initiative was launched in the USA, the new name, "nanoparticle," had become more common (for example, the same senior author's paper 20 years later addressing the same issue, lognormal distribution of sizes). Nanoparticles can exhibit size-related properties significantly different from those of either fine particles or bulk materials.

Nanoclusters have at least one dimension between 1 and 10 nanometers and a narrow size distribution. Nanopowders are agglomerates of ultrafine particles, nanoparticles, or nanoclusters. Nanometer-sized single crystals, or single-domain ultrafine particles, are often referred to as nanocrystals.

Background

Although nanoparticles are associated with modern science, they have a long history. Nanoparticles were used by artisans as far back as Rome in the fourth century in the famous Lycurgus cup made of dichroic glass as well as the ninth century in Mesopotamia for creating a glittering effect on the surface of pots. In modern times, pottery from the Middle Ages and Renaissance often retains a distinct gold- or copper-colored metallic glitter. This luster is caused by a metallic film that was applied to the transparent surface of a glazing. The luster can still be visible if the film has resisted atmospheric oxidation and other weathering.

The luster originates within the film itself, which contains silver and copper nanoparticles dispersed homogeneously in the glassy matrix of the ceramic glaze. These nanoparticles are created by the artisans by adding copper and silver salts and oxides together with vinegar, ochre, and clay on the surface of previously-glazed pottery. The object is then placed into a kiln and heated to about 600 °C in a reducing atmosphere. In heat the glaze softens, causing the copper and silver ions to migrate into the outer layers of the glaze. There the reducing atmosphere reduced the ions back to metals, which then came together forming the nanoparticles that give the color and optical effects. Luster technique showed that ancient craftsmen had a sophisticated empirical knowledge of materials. The technique originated in the Muslim world. As Muslims were not allowed to use gold in artistic representations, they sought a way to create a similar effect without using real gold. The solution they found was using luster.

Michael Faraday provided the first description, in scientific terms, of the optical properties of nanometer-scale metals in his classic 1857 paper. In a subsequent paper, the author (Turner) points out that: "It is well known that when thin leaves of gold or silver are mounted upon glass and heated to a temperature that is well below a red heat (~500 °C), a remarkable change of properties takes place, whereby the continuity of the metallic film is destroyed. The result is that white light is now freely transmitted, reflection is correspondingly diminished, while the electrical resistivity is enormously increased."

Uniformity

The chemical processing and synthesis of high-performance technological components for the private, industrial, and military sectors requires the use of high-purity ceramics (oxide ceramics, such as aluminium oxide or copper(II) oxide), polymers, glass-ceramics, and composite materials, as metal carbides (SiC), nitrides (Aluminum nitrides, Silicon nitride), metals (Al, Cu), non-metals (graphite, carbon nanotubes) and layered (Al + Aluminium carbonate, Cu + C). In condensed bodies formed from fine powders, the irregular particle sizes and shapes in a typical powder often lead to non-uniform packing morphologies that result in packing density variations in the powder compact.

Uncontrolled agglomeration of powders due to attractive van der Waals forces can also give rise to microstructural heterogeneity. Differential stresses that develop as a result of non-uniform drying shrinkage are directly related to the rate at which the solvent can be removed, and thus highly dependent upon the distribution of porosity. Such stresses have been associated with a plastic-to-brittle transition in consolidated bodies, and can yield to crack propagation in the unfired body if not relieved.

In addition, any fluctuations in packing density in the compact as it is prepared for the kiln are often amplified during the sintering process, yielding inhomogeneous densification. Some pores and other structural defects associated with density variations have been shown to play a detrimental role in the sintering process by growing and thus limiting end-point densities. Differential stresses arising from inhomogeneous densification have also been shown to result in the propagation of internal cracks, thus becoming the strength-controlling flaws.

Inert gas evaporation and inert gas deposition are free many of these defects due to the distillation (cf. purification) nature of the process and having enough time to form single crystal particles, however even their non-aggreated deposits have lognormal size distribution, which is typical with nanoparticles. The reason why modern gas evaporation techniques can produce a relatively narrow size distribution is that aggregation can be avoided. However, even in this case, random residence times in the growth zone, due to the combination of drift and diffusion, result in a size distribution appearing lognormal.

It would, therefore, appear desirable to process a material in such a way that it is physically uniform with regard to the distribution of components and porosity, rather than using particle size distributions that will maximize the green density. The containment of a uniformly dispersed assembly of strongly interacting particles in suspension requires total control over interparticle forces. Monodisperse nanoparticles and colloids provide this potential.

Monodisperse powders of colloidal silica, for example, may therefore be stabilized sufficiently to ensure a high degree of order in the colloidal crystal or polycrystalline colloidal solid that results from aggregation. The degree of order appears to be limited by the time and space allowed for longer-range correlations to be established. Such defective polycrystalline colloidal structures would appear to be the basic elements of submicrometer colloidal materials science and, therefore, provide the first step in developing a more rigorous understanding of the mechanisms involved in microstructural evolution in high performance materials and components.

Properties

Nanoparticles are of great scientific interest as they are, in effect, a bridge between bulk materials and atomic or molecular structures. A bulk material should have constant physical properties regardless of its size, but at the nano-scale size-dependent properties are often observed. Thus, the properties of materials change as their size approaches the nanoscale and as the percentage of the surface in relation to the percentage of the volume of a material becomes significant. For bulk materials larger than one micrometer (or micron), the percentage of the surface is insignificant in relation to the volume in the bulk of the material. *The interesting and sometimes unexpected properties of nanoparticles are therefore largely due to the large surface area of the material, which dominates the contributions made by the small bulk of the material.*

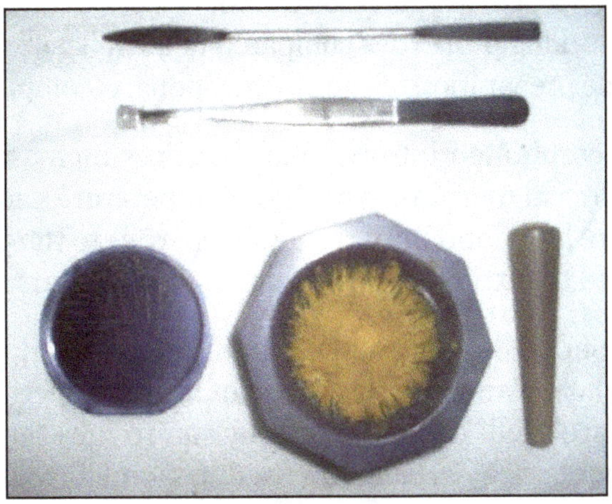

Silicon nanopowder.

1 kg of particles of 1 mm^3 has the same surface area as 1 mg of particles of 1 nm^3.

Nanoparticles often possess unexpected optical properties as they are small enough to confine their electrons and produce quantum effects. For example, gold nanoparticles appear deep-red to black in solution. Nanoparticles of yellow gold and grey silicon are red in color. Gold nanoparticles melt at much lower temperatures (~300 °C for 2.5 nm size) than the gold slabs (1064 °C);. Absorption of solar radiation is much higher in materials composed of nanoparticles than it is in thin films of continuous sheets of material. In both solar PV and solar thermal applications, controlling the size, shape, and material of the particles, it is possible to control solar absorption.

Other size-dependent property changes include quantum confinement in semiconductor particles, surface plasmon resonance in some metal particles and superparamagnetism in magnetic materials. What would appear ironic is that the changes in physical properties are not always desirable. Ferromagnetic materials smaller than 10 nm can switch their magnetisation direction using room temperature thermal energy, thus making them unsuitable for memory storage.

Suspensions of nanoparticles are possible since the interaction of the particle surface with the solvent is strong enough to overcome density differences, which otherwise usually result in a material either sinking or floating in a liquid.

The high surface area to volume ratio of nanoparticles provides a tremendous driving force for diffusion, especially at elevated temperatures. Sintering can take place at lower temperatures, over shorter time scales than for larger particles. In theory, this does not affect the density of the final product, though flow difficulties and the tendency of nanoparticles to agglomerate complicates matters. Moreover, nanoparticles have been found to impart some extra properties to various day to day products. For example, the presence of titanium dioxide nanoparticles imparts what we call the self-cleaning effect, and, the size being nano-range, the particles cannot be observed. Zinc oxide particles have been found to have superior UV blocking properties compared to its bulk substitute. This is one of the reasons why it is often used in the preparation of sunscreen lotions, is completely photostable and toxic.

Clay nanoparticles when incorporated into polymer matrices increase reinforcement, leading to stronger plastics, verifiable by a higher glass transition temperature and other mechanical property tests. These nanoparticles are hard, and impart their properties to the polymer (plastic). Nanoparticles have also been attached to textile fibers in order to create smart and functional clothing.

Metal, dielectric, and semiconductor nanoparticles have been formed, as well as hybrid structures (e.g., core–shell nanoparticles). Nanoparticles made of semiconducting material may also be labeled quantum dots if they are small enough (typically sub 10 nm) that quantization of electronic energy levels occurs. Such nanoscale particles are used in biomedical applications as drug carriers or imaging agents.

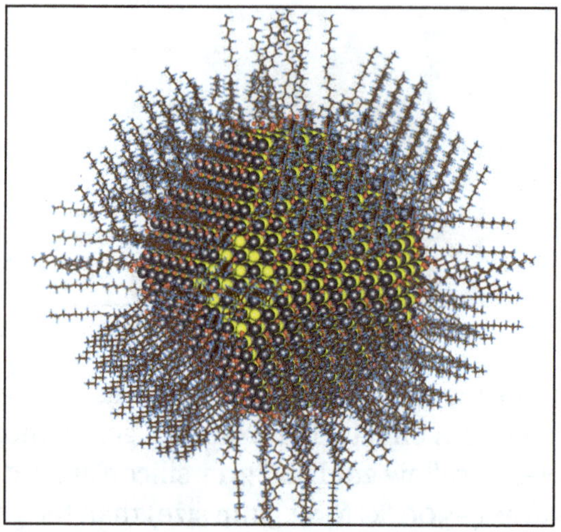

Semiconductor nanoparticle (quantum dot) of lead sulfide with complete passivation by oleic acid, oleyl amine and hydroxyl ligands (size ~5nm).

Semi-solid and soft nanoparticles have been manufactured. A prototype nanoparticle of semi-solid nature is the liposome. Various types of liposome nanoparticles are currently used clinically as delivery systems for anticancer drugs and vaccines.

Nanoparticles with one half hydrophilic and the other half hydrophobic are termed Janus particles and are particularly effective for stabilizing emulsions. They can self-assemble at water/oil interfaces and act as solid surfactants.

Hydrogel nanoparticles made of N-isopropylacrylamide hydrogel core shell can be dyed with affinity baits, internally. These affinity baits allow the nanoparticles to isolate and remove undesirable proteins while enhancing the target analytes.

Synthesis

There are several methods for creating nanoparticles, including gas condensation, attrition, chemical precipitation, pyrolysis and hydrothermal synthesis. In attrition, macro- or micro-scale particles are ground in a ball mill, a planetary ball mill, or other size-reducing mechanism. The resulting particles are air classified to recover nanoparticles. In pyrolysis, a vaporous precursor (liquid or gas) is forced through an orifice at high pressure and burned. The resulting solid (a version of soot) is air classified to recover oxide particles from by-product gases. Traditional pyrolysis often results in aggregates and agglomerates rather than single primary particles. Ultrasonic nozzle spray pyrolysis (USP) on the other hand aids in preventing agglomerates from forming.

A thermal plasma can deliver the energy to vaporize small micrometer-size particles. The thermal plasma temperatures are in the order of 10,000 K, so that solid powder easily evaporates. Nanoparticles are formed upon cooling while exiting the plasma region. The main types of the thermal plasma torches used to produce nanoparticles are dc plasma jet, dc arc plasma, and radio frequency (RF) induction plasmas. In the arc plasma reactors, the energy necessary for evaporation and reaction is provided by an electric arc formed between the anode and the cathode. For example, silica sand can be vaporized with an arc plasma at atmospheric pressure, or thin aluminum wires can be vaporized by exploding wire method. The resulting mixture of plasma gas and silica vapour can be rapidly cooled by quenching with oxygen, thus ensuring the quality of the fumed silica produced.

In RF induction plasma torches, energy coupling to the plasma is accomplished through the electromagnetic field generated by the induction coil. The plasma gas does not come in contact with electrodes, thus eliminating possible sources of contamination and allowing the operation of such plasma torches with a wide range of gases including inert, reducing, oxidizing, and other corrosive atmospheres. The working frequency is typically between 200 kHz and 40 MHz. Laboratory units run at power levels in the order of 30–50 kW, whereas the large-scale industrial units have been tested at power levels up to 1 MW. As the residence time of the injected feed droplets in the plasma is very short, it is important that the droplet sizes are small enough in order to obtain complete evaporation. The RF plasma method has been used to synthesize different nanoparticle materials, for example synthesis of various ceramic nanoparticles such as oxides, carbours/carbides, and nitrides of Ti and Si.

Inert-gas condensation is frequently used to make nanoparticles from metals with low melting points. The metal is vaporized in a vacuum chamber and then supercooled with an inert gas stream. The supercooled metal vapor condenses into nanometer-size particles, which can be entrained in the inert gas stream and deposited on a substrate or studied in situ.

Nanoparticles can also be formed using radiation chemistry. Radiolysis from gamma rays can create strongly active free radicals in solution. This relatively simple technique uses a minimum number of chemicals. These including water, a soluble metallic salt, a radical scavenger (often a secondary alcohol), and a surfactant (organic capping agent). High gamma doses on the order of 10^4 Gray are required. In this process, reducing radicals will drop metallic ions down to the zero-valence state. A scavenger chemical will preferentially interact with oxidizing radicals to prevent the re-oxidation of the metal. Once in the zero-valence state, metal atoms begin to coalesce into particles. A chemical surfactant surrounds the particle during formation and regulates its growth. In sufficient concentrations, the surfactant molecules stay attached to the particle. This prevents it from dissociating or forming clusters with other particles. Formation of nanoparticles using the radiolysis method allows for tailoring of particle size and shape by adjusting precursor concentrations and gamma dose.

Sol-gel

The sol-gel process is a wet-chemical technique (also known as chemical solution deposition) widely used recently in the fields of materials science and ceramic engineering. Such methods are used primarily for the fabrication of materials (typically a metal oxide) starting from a chemical solution (*sol*, short for solution), which acts as the precursor for an integrated network (or *gel*) of either discrete particles or network polymers.

Typical precursors are metal alkoxides and metal chlorides, which undergo hydrolysis and polycondensation reactions to form either a network "elastic solid" or a colloidal suspension (or dispersion) – a system composed of discrete (often amorphous) submicrometer particles dispersed to various degrees in a host fluid. Formation of a metal oxide involves connecting the metal centers with oxo (M-O-M) or hydroxo (M-OH-M) bridges, therefore generating metal-oxo or metal-hydroxo polymers in solution. Thus, the sol evolves toward the formation of a gel-like diphasic system containing both a liquid phase and solid phase whose morphologies range from discrete particles to continuous polymer networks.

In the case of the colloid, the volume fraction of particles (or particle density) may be so low that a significant amount of fluid may need to be removed initially for the gel-like properties to be recognized. This can be accomplished in a number of ways. The most simple method is to allow time for sedimentation to occur, and then pour off the remaining liquid. Centrifugation can also be used to accelerate the process of phase separation.

Removal of the remaining liquid (solvent) phase requires a drying process, which typically causes shrinkage and densification. The rate at which the solvent can be removed is ultimately determined by the distribution of porosity in the gel. The ultimate microstructure of the final component will clearly be strongly influenced by changes implemented during this phase of processing. Afterward, a thermal treatment, or firing process, is often necessary in order to favor further polycondensation and enhance mechanical properties and structural stability via final sintering, densification, and grain growth. One of the distinct advantages of using this methodology as opposed to the more traditional processing techniques is that densification is often achieved at a much lower temperature.

The precursor sol can be either deposited on a substrate to form a film (e.g., by dip-coating or spin-coating), cast into a suitable container with the desired shape (e.g., to obtain a monolithic ceramics, glasses, fibers, membranes, aerogels), or used to synthesize powders (e.g., microspheres, nanospheres). The sol-gel approach is a cheap and low-temperature technique that allows for the fine control of the product's chemical composition. Even small quantities of dopants, such as organic dyes and rare earth metals, can be introduced in the sol and end up uniformly dispersed in the final product. It can be used in ceramics processing and manufacturing as an investment casting material, or as a means of producing very thin films of metal oxides for various purposes. Sol-gel derived materials have diverse applications in optics, electronics, energy, space, (bio)sensors, medicine (e.g., controlled drug release) and separation (e.g., chromatography) technology.

Colloids

The term colloid is used primarily to describe a range of mixtures which have solid particles dispersed in a liquid medium. The term applies only if the particles are larger than atomic dimensions but small enough to exhibit Brownian motion. If the particles are large enough, their dynamic behavior in any given period of time in suspension would be governed by forces of gravity and sedimentation. If the particles are small enough, their irregular motion in suspension can be attributed to the collective bombardment of a myriad of thermally agitated molecules in the liquid suspending medium, as described originally by Albert Einstein in his dissertation. Einstein showed evidence that water was made up of discrete molecules by characterizing the observed erratic particle behavior using the theory of Brownian motion, with sedimentation being a possible result. The critical size range (or particle diameter) typically ranges from nanometers (10^{-9} m) to micrometers (10^{-6} m).

Morphology

Nanostars of vanadium(IV) oxide.

Scientists have taken to naming their particles after the real-world shapes that they might represent. Nanospheres, nanochains, nanoreefs, nanoboxes and more have appeared in the literature. These morphologies sometimes arise spontaneously as an effect of a templating or directing agent present in the synthesis such as miscellar emulsions or anodized alumina pores, or from the innate crystallographic growth patterns of the materials themselves. Some of these morphologies may serve a purpose, such as long carbon nanotubes used to bridge an electrical junction, or just a scientific curiosity like the stars shown.

Amorphous particles usually adopt a spherical shape (due to their microstructural isotropy), whereas the shape of anisotropic microcrystalline whiskers corresponds to their particular crystal habit. At the small end of the size range, nanoparticles are often referred to as clusters. Spheres, rods, fibers, and cups are just a few of the shapes that have been grown. The study of fine particles is called micromeritics.

Characterization

The majority of nanoparticle characterization techniques are light-based, but a non-optical nanoparticle characterization technique called Tunable Resistive Pulse Sensing (TRPS) has been developed that enables the simultaneous measurement of size, concentration and surface charge for a wide variety of nanoparticles. This technique, which applies the Coulter Principle, allows for particle-by-particle quantification of these three nanoparticle characteristics with high resolution.

Functionalization

The surface coating of nanoparticles determines many of their properties, notably stability, solubility, and targeting. A coating that is multivalent or polymeric confers high stability. Functionalized nanomaterial-based catalysts can be used for catalysis of many known organic reactions.

Surface Coating for Biological Applications

For biological applications, the surface coating should be polar to give high aqueous solubility and prevent nanoparticle aggregation. In serum or on the cell surface, highly charged coatings promote non-specific binding, whereas polyethylene glycol linked to terminal hydroxyl or methoxy groups repel non-specific interactions. Nanoparticles can be linked to biological molecules that can act as address tags, to direct the nanoparticles to specific sites within the body, specific organelles within the cell, or to follow specifically the movement of individual protein or RNA molecules in living cells. Common address tags are monoclonal antibodies, aptamers, streptavidin or peptides. These targeting agents should ideally be covalently linked to the nanoparticle and should be present in a controlled number per nanoparticle. Multivalent nanoparticles, bearing multiple targeting groups, can cluster receptors, which can activate cellular signaling pathways, and give stronger anchoring. Monovalent nanoparticles, bearing a single binding site, avoid clustering and so are preferable for tracking the behavior of individual proteins.

Red blood cell coatings can help nanoparticles evade the immune system.

Health and Safety

Nanoparticles present possible dangers, both medically and environmentally. Most of these are due to the high surface to volume ratio, which can make the particles very reactive or catalytic. They are also able to pass through cell membranes in organisms, and their interactions with biological systems are relatively unknown. However, it is unlikely the particles would enter the cell nucleus, Golgi complex, endoplasmic reticulum or other internal cellular components due to the particle size and intercellular agglomeration. A recent study looking at the effects of ZnO nanoparticles on human immune cells has found varying levels of susceptibility to cytotoxicity. There are concerns that pharmaceutical companies, seeking regulatory approval for nano-reformulations of existing medicines, are relying on safety data produced during clinical studies of the earlier, pre-reformulation version of the medicine. This could result in regulatory bodies, such as the FDA, missing new side effects that are specific to the nano-reformulation.

Whether cosmetics and sunscreens containing nanomaterials pose health risks remains largely unknown at this stage. However considerable research has demonstrated that zinc nanoparticles are not absorbed into the bloodstream in vivo.

Concern has also been raised over the health effects of respirable nanoparticles from certain combustion processes. As of 2013 the U.S. Environmental Protection Agency was investigating the safety of the following nanoparticles:

- Carbon Nanotubes: Carbon materials have a wide range of uses, ranging from composites for use in vehicles and sports equipment to integrated circuits for electronic components. The interactions between nanomaterials such as carbon nanotubes and natural organic matter strongly influence both their aggregation and deposition, which strongly affects their transport, transformation, and exposure in aquatic environments. In past research, carbon nanotubes exhibited some toxicological impacts that will be evaluated in various environmental settings in current EPA chemical safety research. EPA research will provide data, models, test methods, and best practices to discover the acute health effects of carbon nanotubes and identify methods to predict them.
- Cerium oxide: Nanoscale cerium oxide is used in electronics, biomedical supplies, energy, and fuel additives. Many applications of engineered cerium oxide nanoparticles naturally disperse themselves into the environment, which increases the risk of exposure. There is ongoing exposure to new diesel emissions using fuel additives containing CeO_2 nanoparticles, and the environmental and public health impacts of this new technology are unknown. EPA's chemical safety research is assessing the environmental, ecological, and health implications of nanotechnology-enabled diesel fuel additives.
- Titanium dioxide: Nano titanium dioxide is currently used in many products. Depending on the type of particle, it may be found in sunscreens, cosmetics, and paints and coatings. It is also being investigated for use in removing contaminants from drinking water.
- Nano Silver: Nano silver is being incorporated into textiles, clothing, food packaging, and other materials to eliminate bacteria. EPA and the U.S. Consumer Product Safety Commission are studying certain products to see whether they transfer nano-size silver particles in real-world scenarios. EPA is researching this topic to better understand how much nano-silver children come in contact with in their environments.
- Iron: While nano-scale iron is being investigated for many uses, including "smart fluids" for uses such as optics polishing and as a better-absorbed iron nutrient supplement, one of its more prominent current uses is to remove contamination from groundwater. This use, supported by EPA research, is being piloted at a number of sites across the country.

Laser Applications

The use of nanoparticles in laser dye-doped poly(methyl methacrylate) (PMMA) laser gain media was demonstrated in 2003 and it has been shown to improve conversion efficiencies and to decrease laser beam divergence. Researchers attribute the reduction in beam divergence to improved dn/dT characteristics of the organic-inorganic dye-doped nanocomposite. The optimum composition reported by these researchers is 30% w/w of SiO_2 (~ 12 nm) in dye-doped PMMA.

Medicinal Applications

- Liposome.
- Dendrimer.
- Iron oxide nanoparticles.
- Nanomedicine.
- Polymer-drug conjugate.
- Polymeric nanoparticle.

Magnetic Nanoparticles

Magnetic nanoparticles are a class of nanoparticle that can be manipulated using magnetic fields. Such particles commonly consist of two components, a magnetic material, often iron, nickel and cobalt, and a chemical component that has functionality. While nanoparticles are smaller than 1 micrometer in diameter (typically 5–500 nanometers), the larger microbeads are 0.5–500 micrometer in diameter. Magnetic nanoparticle clusters that are composed of a number of individual magnetic nanoparticles are known as magnetic nanobeads with a diameter of 50–200 nanometers. Magnetic nanoparticle clusters are a basis for their further magnetic assembly into magnetic nanochains. The magnetic nanoparticles have been the focus of much research recently because they possess attractive properties which could see potential use in catalysis including nanomaterial-based catalysts, biomedicine and tissue specific targeting, magnetically tunable colloidal photonic crystals, microfluidics, magnetic resonance imaging, magnetic particle imaging, data storage, environmental remediation, nanofluids, and optical filters, defect sensor and cation sensors.

Properties

The physical and chemical properties of magnetic nanoparticles largely depend on the synthesis method and chemical structure. In most cases, the particles range from 1 to 100 nm in size and may display superparamagnetism.

Types of Magnetic Nanoparticles

Oxides: Ferrites

Ferrite nanoparticles or iron oxide nanoparticles (iron oxides in crystal structure of maghemite or magnetite) are the most explored magnetic nanoparticles up to date. Once the ferrite particles become smaller than 128 nm they become superparamagnetic which prevents self agglomeration since they exhibit their magnetic behavior only when an external magnetic field is applied. The magnetic moment of ferrite nanoparticles can be greatly increased by controlled clustering of a number of individual superparamagnetic nanoparticles into superparamagnetic nanoparticle clusters, namely magnetic nanobeads. With the external magnetic field switched off, the remanence falls back to zero. Just like non-magnetic oxide nanoparticles, the surface of ferrite nanoparticles is often modified by surfactants, silica, silicones or phosphoric acid derivatives to increase their stability in solution.

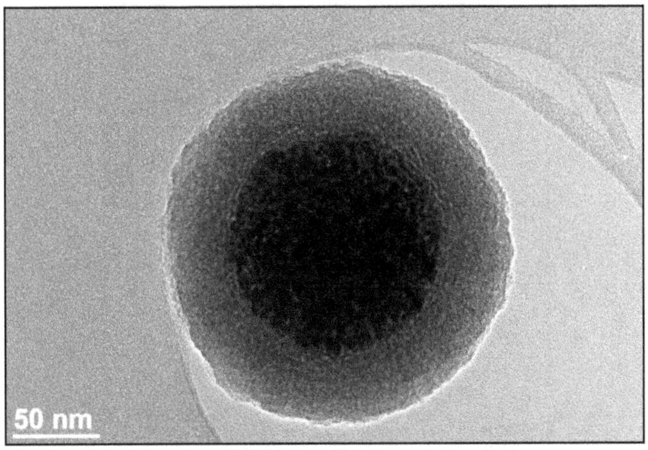

TEM image of a maghemite magnetic nanoparticle cluster with silica shell.

Ferrites with a Shell

The surface of a maghemite or magnetite magnetic nanoparticle is relatively inert and does not usually allow strong covalent bonds with functionalization molecules. However, the reactivity of the magnetic nanoparticles can be improved by coating a layer of silica onto their surface. The silica shell can be easily modified with various surface functional groups via covalent bonds between organo-silane molecules and silica shell. In addition, some fluorescent dye molecules can be covalently bonded to the functionalized silica shell.

Ferrite nanoparticle clusters with narrow size distribution consisting of superparamagnetic oxide nanoparticles (~ 80 maghemite superparamagnetic nanoparticles per bead) coated with a silica shell have several advantages over metallic nanoparticles:

- Higher chemical stability (crucial for biomedical applications).
- Narrow size distribution (crucial for biomedical applications).
- Higher colloidal stability since they do not magnetically agglomerate.
- Magnetic moment can be tuned with the nanoparticle cluster size.
- Retained superparamagnetic properties (independent of the nanoparticle cluster size).
- Silica surface enables straightforward covalent functionalization.

Metallic

Metallic nanoparticles may be beneficial for some technical applications due to their higher magnetic moment whereas oxides (maghemite, magnetite) would be beneficial for biomedical applications. This also implies that for the same moment, metallic nanoparticles can be made smaller than their oxide counterparts. On the other hand, metallic nanoparticles have the great disadvantage of being pyrophoric and reactive to oxidizing agents to various degrees. This makes their handling difficult and enables unwanted side reactions which makes them less appropriate for biomedical applications. Colloid formation for metallic particles is also much more challenging.

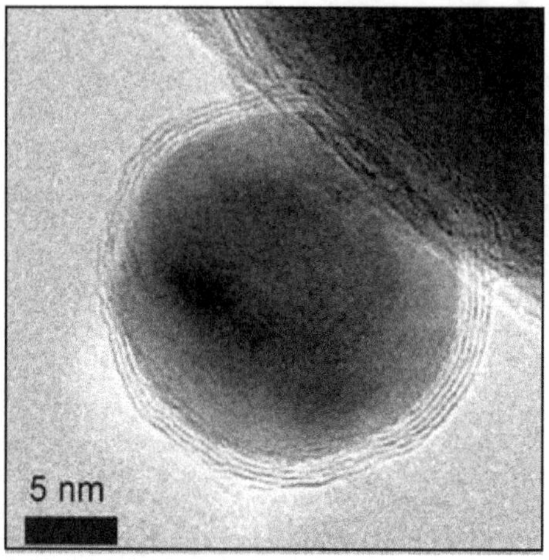

Cobalt nanoparticle with graphene shell (note: The individual graphene layers are visible).

Metallic with a Shell

The metallic core of magnetic nanoparticles may be passivated by gentle oxidation, surfactants, polymers and precious metals. In an oxygen environment, Co nanoparticles form an anti-ferromagnetic CoO layer on the surface of the Co nanoparticle. Recently, work has explored the synthesis and exchange bias effect in these Co core CoO shell nanoparticles with a gold outer shell. Nanoparticles with a magnetic core consisting either of elementary Iron or Cobalt with a nonreactive shell made of graphene have been synthesized recently. The advantages compared to ferrite or elemental nanoparticles are:

- Higher magnetization.
- Higher stability in acidic and basic solution as well as organic solvents
- Chemistry on the graphene surface via methods already known for carbon nanotubes.

Synthesis

Several methods exist for preparing magnetic nanoparticle.

Co-precipitation

Co-precipitation is a facile and convenient way to synthesize iron oxides (either Fe_3O_4 or γ-Fe_2O_3) from aqueous Fe^{2+}/Fe^{3+} salt solutions by the addition of a base under inert atmosphere at room temperature or at elevated temperature. The size, shape, and composition of the magnetic nanoparticles very much depends on the type of salts used (e.g.chlorides, sulfates, nitrates), the Fe^{2+}/Fe^{3+} ratio, the reaction temperature, the pH value and ionic strength of the media, and the mixing rate with the base solution used to provoke the precipitation. The co-precipitation approach has been used extensively to produce ferrite nanoparticles of controlled sizes and magnetic properties. A variety of experimental arrangements have been reported to facilitate continuous and large–scale co–precipitation of magnetic particles by rapid mixing. Recently, the growth rate of the magnetic nanoparticles was measured in real-time during the precipitation of magnetite nanoparticles by an integrated AC magnetic susceptometer within the mixing zone of the reactants.

Thermal Decomposition

Magnetic nanocrystals with smaller size can essentially be synthesized through the thermal decomposition of alkaline organometallic compounds in high-boiling organic solvents containing stabilizing surfactants.

Microemulsion

Using the microemulsion technique, metallic cobalt, cobalt/platinum alloys, and gold-coated cobalt/platinum nanoparticles have been synthesized in reverse micelles of cetyltrimethlyammonium bromide, using 1-butanol as the cosurfactant and octane as the oil phase.

Flame Spray Synthesis

Using flame spray pyrolysis and varying the reaction conditions, oxides, metal or carbon coated nanoparticles are produced at a rate of > 30 g/h.

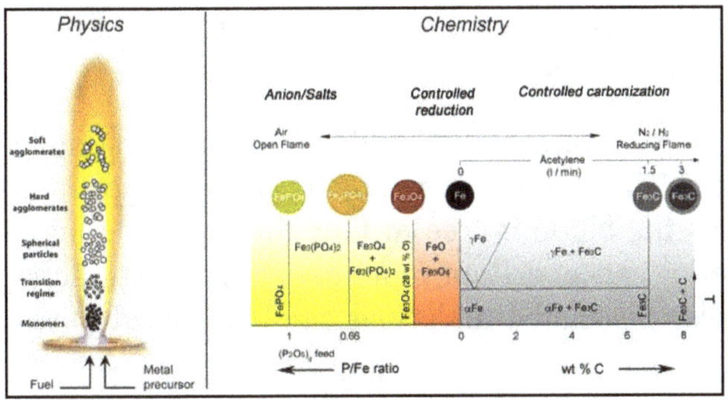

Various flame spray conditions and their impact on the resulting nanoparticles.

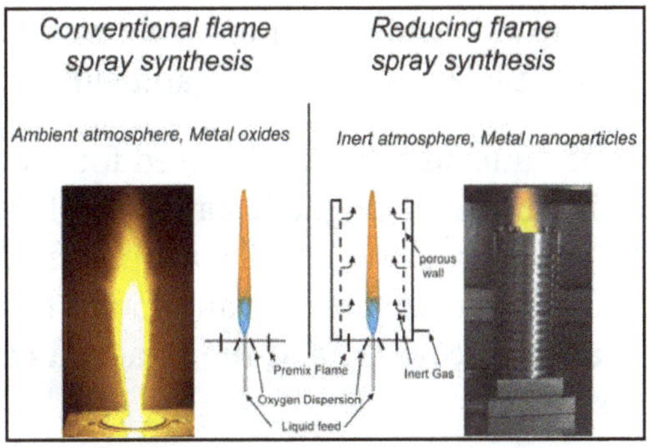

Operational layout differences between conventional and reducing flame spray synthesis.

Potential Applications

A wide variety of potential applications have been envisaged. Since magnetic nanoparticles are expensive to produce, there is interest in their recycling or for highly specialized applications. They are only used in scientific research. An industrial use has yet to be established.

The potential and versatility of magnetic chemistry arises from the fast and easy separation of the magnetic nanoparticles, eliminating tedious and costly separation processes usually applied in chemistry. Furthermore, the magnetic nanoparticles can be guided via a magnetic field to the desired location which could, for example, enable pinpoint precision in fighting cancer.

Medical Diagnostics and Treatments

Magnetic nanoparticles have been examined for use in an experimental cancer treatment called magnetic hyperthermia in which an alternating magnetic field (AMF) is used to heat the nanoparticles. To achieve sufficient magnetic nanoparticle heating, the AMF typically has a frequency between 100–500 kHz, although significant research has been done at lower frequencies as well as frequencies as high as 10 MHz, with the amplitude of the field usually between 8-16kAm^{-1}.

Affinity ligands such as epidermal growth factor (EGF), folic acid, aptamers, lectins etc. can be attached to the magnetic nanoparticle surface with the use of various chemistries. This enables targeting of magnetic nanoparticles to specific tissues or cells. This strategy is used in cancer research to target and treat tumors in combination with magnetic hyperthermia or nanoparticle-delivered cancer drugs. Despite research efforts, however, the accumulation of nanoparticles inside of cancer tumors of all types is sub-optimal, even with affinity ligands. Willhelm et al. conducted a broad analysis of nanoparticle delivery to tumors and concluded that the median amount of injected dose reaching a solid tumor is only 0.7%. The challenge of accumulating large amounts of nanoparticles inside of tumors is arguably the biggest obstacle facing nanomedicine in general. While direct injection is used in some cases, intravenous injection is most often preferred to obtain a good distribution of particles throughout the tumor. Magnetic nanoparticles have a distinct advantage in that they can accumulate in desired regions via magnetically guided delivery, although this technique still needs further development to achieve optimal delivery to solid tumors.

Another potential treatment of cancer includes attaching magnetic nanoparticles to free-floating cancer cells, allowing them to be captured and carried out of the body. The treatment has been tested in the laboratory on mice and will be looked at in survival studies.

Magnetic nanoparticles can be used for the detection of cancer. Blood can be inserted onto a microfluidic chip with magnetic nanoparticles in it. These magnetic nanoparticles are trapped inside due to an externally applied magnetic field as the blood is free to flow through. The magnetic nanoparticles are coated with antibodies targeting cancer cells or proteins. The magnetic nanoparticles can be recovered and the attached cancer-associated molecules can be assayed to test for their existence.

Magnetic nanoparticles can be conjugated with carbohydrates and used for detection of bacteria. Iron oxide particles have been used for the detection of Gram negative bacteria like Escherichia coli and for detection of Gram positive bacteria like Streptococcus suis.

Magnetic Immunoassay

Magnetic immunoassay (MIA) is a novel type of diagnostic immunoassay utilizing magnetic nanobeads as labels in lieu of conventional, enzymes, radioisotopes or fluorescent moieties. This assay involves the specific binding of an antibody to its antigen, where a magnetic label is conjugated to one element of the pair. The presence of magnetic nanobeads is then detected by a magnetic reader (magnetometer) which measures the magnetic field change induced by the beads. The signal measured by the magnetometer is proportional to the analyte (virus, toxin, bacteria, cardiac marker,etc.) quantity in the initial sample.

Waste Water Treatment

Thanks to the easy separation by applying a magnetic field and the very large surface to volume ratio, magnetic nanoparticles have a potential for treatment of contaminated water. In this method, attachment of EDTA-like chelators to carbon coated metal nanomagnets results in a magnetic reagent for the rapid removal of heavy metals from solutions or contaminated water by three orders of magnitude to concentrations as low as micrograms per Litre. Magnetic nanobeads or nanoparticle clusters composed of FDA-approved oxide superparamagnetic nanoparticles (e.g. maghemite, magnetite) hold much potential for waste water treatment since they express excellent biocompatibility which concerning the environmental impacts of the material is an advantage compared to metallic nanoparticles.

Supported Enzymes and Peptides

Enzymes, proteins, and other biologically and chemically active substances have been immobilized on magnetic nanoparticles. They are of interest as possible supports for solid phase synthesis.

This technology is potentially relevant to cellular labelling/cell separation, detoxification of biological fluids, tissue repair, drug delivery, magnetic resonance imaging, hyperthermia and magnetofection.

Catalyst Support

Magnetic nanoparticles are of potential use as a catalyst or catalyst supports. In chemistry, a catalyst support is the material, usually a solid with a high surface area, to which a catalyst is affixed. The reactivity of heterogeneous catalysts occurs at the surface atoms. Consequently, great effort is made to maximize the surface area of a catalyst by distributing it over the support. The support may be inert or participate in the catalytic reactions. Typical supports include various kinds of carbon, alumina, and silica. Immobilizing the catalytic center on top of nanoparticles with a large surface to volume ratio addresses this problem. In the case of magnetic nanoparticles it adds the property of facile a separation. An early example involved a rhodium catalysis attached to magnetic nanoparticles.

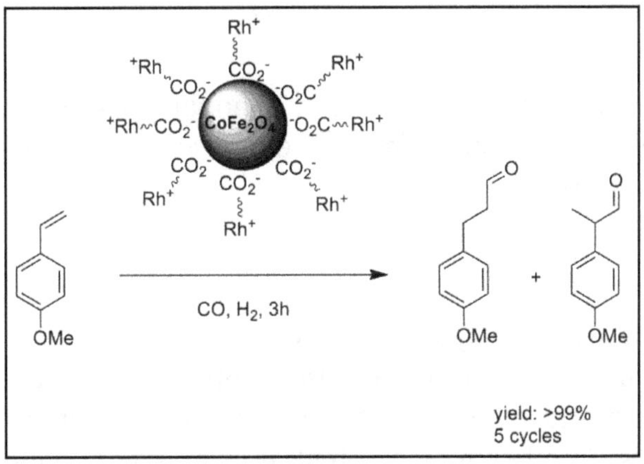

In another example, the stable radical TEMPO was attached to the graphene-coated cobalt nanoparticles via a diazonium reaction. The resulting catalyst was then used for the chemoselective oxidation of primary and secondary alcohols.

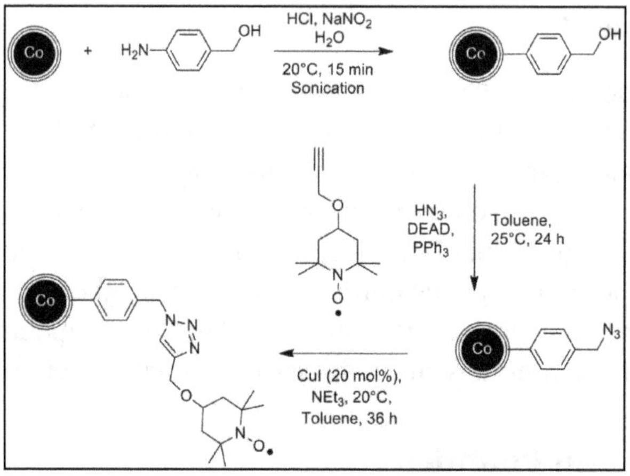

The catalytic reaction can be conducted in a continuous flow reactor instead of a batch reactor with no remains of the catalyst in the end product. Graphene coated cobalt nanoparticles have been used for that experiment since they exhibit a higher magnetization than Ferrite nanoparticles, which is essential for a fast and clean separation via external magnetic field.

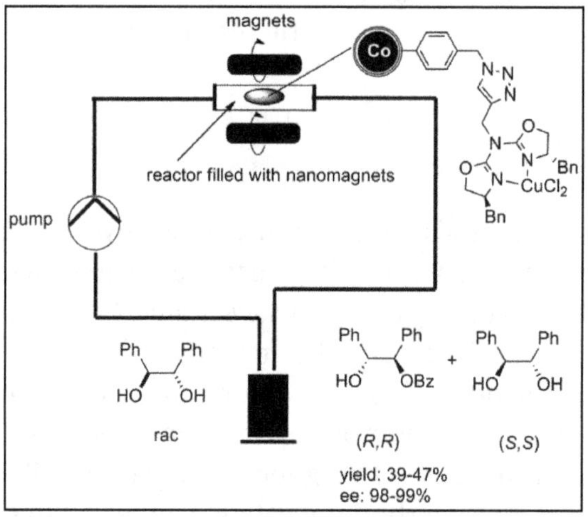

Biomedical Imaging

There are many applications for iron-oxide based nanoparticles in concert with magnetic resonance imaging. Magnetic CoPt nanoparticles are being used as an MRI contrast agent for transplanted neural stem cell detection.

Cancer Therapy

In magnetic fluid hyperthermia, nanoparticles of different types like Iron oxide, magnetite, maghemite or even gold are injected in tumor and then subjected under a high frequency magnetic field. These nanoparticles produce heat that typically increases tumor temperature to 40-46 °C, which can kill cancer cells. Another major potential of magnetic nanoparticles is the ability to combine heat (hyperthermia) and drug release for a cancer treatment. Numerous studies have shown particle constructs that can be loaded with a drug cargo and magnetic nanoparticles. The most prevalent construct is the “Magnetoliposome”, which is a liposome with magnetic nanoparticles typically embedded in the lipid bilayer. Under an alternating magnetic field, the magnetic nanoparticles are heated, and this heat permeabilizes the membrane. This causes release of the loaded drug. This treatment option has a lot of potential as the combination of hyperthermia and drug release is likely to treat tumors better than either option alone, but it is still under development.

Information Storage

A promising candidate for high-density storage is the face-centered tetragonal phase FePt alloy. Grain sizes can be as small as 3 nanometers. If it's possible to modify the MNPs at this small scale, the information density that can be achieved with this media could easily surpass 1 Terabyte per square inch.

Genetic Engineering

Magnetic nanoparticles can be used for a variety of genetics applications. One application is the rapid isolation of mRNA. In one application, the magnetic bead is attached to a poly T tail. When mixed with mRNA, the poly A tail of the mRNA will attach to the bead's poly T tail and the isolation takes place simply by placing a magnet on the side of the tube and pouring out the liquid. Magnetic beads have also been used in plasmid assembly. Rapid genetic circuit construction has been achieved by the sequential addition of genes onto a growing genetic chain, using nanobeads as an anchor. This method has been shown to be much faster than previous methods, taking less than an hour to create functional multi-gene constructs in vitro.

Physical Modeling

There are a variety of mathematical models to describe the dynamics of the rotations of magnetic nanoparticles. Simple models include the Langevin function and the Stoner-Wohlfarth model which describe the magnetization of a nanoparticle at equilibrium. The Debye/Rosensz-weig model can be used for low amplitude or high frequency oscillations of particles, which assumes linear response of the magnetization to an oscillating magnetic field. Non-equilibrium approaches include the Langevin equation formalism and the Fokker-Planck equation formalism, and these have been developed extensively to model applications such as magnetic nanoparticle hyperthermia, magnetic nanoparticle imaging (MPI), magnetic spectroscopy and biosensing etc.

Radioactive Nanoparticle

A radioactive nanoparticle is a nanoparticle that contains radioactive materials. Radioactive nanoparticles have applications in medical diagnostics, medical imaging, toxicokinetics, and environmental health, and are being investigated for applications in nuclear nanomedicine. Radioactive nanoparticles present special challenges in operational health physics and internal dosimetry that are not present for other substances, although existing radiation protection measures and hazard controls for nanoparticles generally apply.

Types and Applications

Engineered

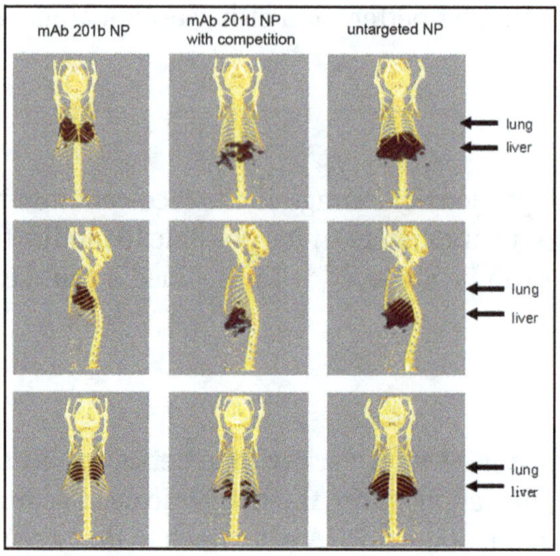

SPECT/CT images of injected gold-coated lanthanum/gadolinium phosphate nanoparticles containing the alpha-emitting radionuclide actinium-225 in a mouse. Depending on the surface functionalization, the particles migrate either to the lungs or the liver.

Engineered radioactive nanoparticles are used in medical imaging techniques such as positron emission tomography and single-photon emission computed tomography, and an aerosol of carbon nanoparticles containing technetium-99m are used in a commercially available procedure for ventilation/perfusion scintigraphy of the lungs. Engineered radioactive nanoparticles are also used as a radiolabel to detect the presence of the nanoparticles themselves in environmental health and toxicokinetics studies.

Engineered radioactive nanoparticles are being investigated for therapeutic use combining nuclear medicine with nanomedicine, especially for cancer. Neutron capture therapy is one such potential application. In addition, nanoparticles can help to sequester the toxic daughter nuclides of alpha emitters when used in radiotherapy.

Nuclear imaging is non-invasive and has high sensitivity, and nanoparticles are useful as a platform for combining multiple copies of targeting vectors and effectors in order to selectively deliver radioisotopes to a specific region of interest. Other benefits of nanoparticles for diagnostic and therapeutic use include increased blood and tumor retention time, as well as the possibility of using their unique physical and chemical properties in treatment. However, the nanoparticles must be engineered to avoid being recognized by the mononuclear phagocyte system and transported to the liver or spleen, often through manipulating their surface functionalization.

Targeting techniques include functionalizing radioactive nanoparticles with antibodies to target them to a specific tissue, and using magnetic nanoparticles that are attracted to a magnet placed over the tumor site. Technetium-99m, indium-111, and iodine-131 are common radioisotopes used for these purposes, with many others used as well. Radioactive nanoparticles can be produced by either synthesizing the nanoparticles directly from the radioactive materials, or by irradiating non-radioactive particles with neutrons or accelerated ions, sometimes *in situ*.

Natural and Incidental

As with all nanoparticles, radioactive nanoparticles can also be naturally occurring or incidentally produced as a byproduct of industrial processes. The main source of naturally occurring nanomaterials containing radionuclides is the decay of radon gas, whose immediate decay products are non-gaseous elements that precipitate into nanoscale particles along with atmospheric dust and vapors. Minor natural sources include primordial radionuclides present in the nanoscale portion of volcanic ash, and primordial and cosmogenic nuclides taken up by plants which are later burned. Radioactive nanoparticles may be incidentally produced by procedures in the nuclear industry such as nuclear reprocessing and the cutting of contaminated objects.

Health and Safety

Radioactive nanoparticles combine the hazards of radioactive materials with the hazards of nanomaterials. Inhalation exposure is the most common route of exposure to airborne particles in the workplace. Animal studies on some classes of nanoparticles indicate pulmonary effects including inflammation, granulomas, and pulmonary fibrosis, which were of similar or greater potency when compared with other known fibrogenic materials such as silica, asbestos, and ultrafine carbon black. Some studies in cells or animals have shown genotoxic or carcinogenic effects, or systemic cadiovascular effects from pulmonary exposure. The hazards of ionizing radiation depend on whether the exposure is acute or chronic, and includes effects like radiation-induced cancer and teratogenesis. In some cases, the inherent physicochemical toxicity of the nanoparticle itself may lead to lower exposure limits than those associated with the radioactivity alone, which is not the case with most radioactive materials.

Radioactive nanoparticles present special challenges in operational health physics and internal dosimetry that are not present for other substances, as the nanoparticles' toxicokinetics depend on their physical and chemical properties including size, shape, and surface chemistry. For example, inhaled nanoparticles will deposit in different locations in the lungs, and will be metabolized and transported through the body differently, than vapors or larger particles. There may also be hazards from associated processes such as strong magnetic fields and cryogens used in imaging equipment, and handling of lab animals in experimental studies. Effective risk assessment and communication is important, as both nanotechnology and radiation have unique considerations with public perception.

Hazard Controls

In general, most elements of a standard radiation protection program are applicable to radioactive nanomaterials, and many hazard controls for nanomaterials will be effective with the radioactive versions. The hierarchy of hazard controls encompasses a succession of five categories of control methods to reduce the risk of illness or injury. The two most effective are elimination and substitution, for example reducing dust exposure by eliminating a sonication process or substituting a nanomaterial slurry or suspension in a liquid solvent instead of a dry powder. Substitutions should consider both the radioactivity and physicochemical hazards of all the options, and also take into account that radioactive nanomaterials are easier to detect than non-radioactive substances.

A fume hood is an engineering control typically used to protect workers using nanoparticles.

Engineering controls should be the primary form of protection, including local exhaust systems such as fume hoods, gloveboxes, biosafety cabinets, and vented balance enclosures; radiation shielding; and access control systems. The need for negative room pressure to prevent contamination of outside areas can conflict with the customary use of positive pressure when pharmaceuticals are being handled, although this can be overcome through use of a cascade pressure system, or by handling nanomaterials in enclosures.

Administrative controls include procedures to limit radiation doses, and contamination control procedures including encouraging good work practices and monitoring for contamination. Personal protective equipment is the least effective and should be used in conjunction with other hazard controls. In general, personal protective equipment intended for radioactive materials should be effective with radioactive nanomaterials, including impervious laboratory coats, goggles, safety gloves, and in some cases respirators, although the greater potential penetration through clothing and mobility in air of nanoparticles should be taken into account.

Nanosponges

Nanosponges are nano sized particles designed to look like red blood cell and protect the body. Nanoparticles have been used for unknown preventative measures, such as silverware and wound care products. Silverware contains small silver nanoparticles which contain an antimicrobial agent. Antimicrobial agents are able to prevent the spread of disease. When silverware was created that was not the intention of the silver product. In more recent years, many wound care products have been made with silver and other various nanoparticles. The universal use of silver in wound care products is mainly for the antimicrobial properties, which helps treat wounds of all severity. They are very minute particles used for a vast majority of things, one including healthcare. A specific type of nanoparticle seen recently in the healthcare field is a nanosponge. Nanosponges are nano sized particles designed to look like red blood cell and protect the body. They are also able to fight and prevent the spread of disease. Nanosponges target toxins, the toxins organize themselves to poke holes inside the red blood cell (nanosponge) membrane this method of poking holes is called pore-forming. After the toxins are attached to the nanosponge, it will lead to cell damage and cell death.

Nanosponges are introduced to the body by injection, and take the form of a red blood cell so that the bacteria or venom attacks it. Once it is attacked, it is trapped within the scaffolding of the nanosponge. After the nanosponge is full of toxins and cannot trap anymore it moves to the liver to filter out the toxins. Researchers are posed with is how to tackle all the different types of bacteria and venom, making a lot of different nanosponges for each specific bacteria and venom is nearly impossible. As of now they are focusing on toxins such as; E. Coli, MRSA, pneumonia, bee venom, snake venom and sea anemone venom. These venoms all use a pore-forming strategy, which is useful when dealing with nanosponges. A single nanosponge can capture many of the bacteria and venoms, instead of being tailored to each individually. Another obstacle researchers are faced with is the lifespan of the nanosponges. Once nanosponges are injected they can move rapidly through the blood system, and be found in the liver to be filtered out within hours. Researchers are working on a technique using hydrogel to coat the nanosponges to increase the life of the nanosponge and help them to remain stationary after injection. What essentially will happen is before the nanosponges are injected they will be coated in a hydrogel that with prolong life, and keep the nanosponges from moving around the body. Having a stationary location for all nanoparticles to reside can more effectively tackle specific regions affected with disease. It was found that 80 percent of nanosponges coated with a hydrogel lasted more than two days after injection. Only 20 percent of nanosponges not coated in the hydrogel lasted two hours after injection, and diffused to other locations in the body.

Nanosponges are known to move rapidly throughout the body, making them efficient in scanning and killing toxins, but irrelevant in tackling a stationary area of disease, or delivering drugs to a specific area. The hydrogel will help immensely in the longevity of the nanosponge, and will allow it to show its full potential in a serious situations. If the nanosponge follows healthcare into the future it can be a tremendous resource for preventative care. Imagine going to the doctor to receive a nanosponge shot to prevent you from acquiring some disease you may be prone to, or possibly a preventative measure before traveling out of the state or country. This technology could save many lives as being used as a preventative measure. Even when patients have to stay at the hospital for long periods of time, maybe due to a surgery or birth. These patients can be injected with nanosponges so they do not acquire Hospital Acquired Infections (HAI). This can reduce the cost to the patient but shortening their stay, and also the hospitals cost for less patients staying longer periods of time, due to a HAI.

Nanofluid

A nanofluid is a fluid containing nanometer-sized particles, called nanoparticles. These fluids are engineered colloidal suspensions of nanoparticles in a base fluid. The nanoparticles used in nanofluids are typically made of metals, oxides, carbides, or carbon nanotubes. Common base fluids include water, ethylene glycol and oil.

Nanofluids have novel properties that make them potentially useful in many applications in heat transfer, including microelectronics, fuel cells, pharmaceutical processes, and hybrid-powered engines, engine cooling/vehicle thermal management, domestic refrigerator, chiller, heat exchanger,in grinding, machining and in boiler flue gas temperature reduction. They exhibit enhanced thermal conductivity and the convective heat transfer coefficient compared to the base fluid. Knowledge of the rheological behaviour of nanofluids is found to be very critical in deciding their suitability for convective heat transfer applications Nanofluids also have special acoustical properties and in ultrasonic fields display additional shear-wave reconversion of an incident compressional wave; the effect becomes more pronounced as concentration increases.

In analysis such as computational fluid dynamics (CFD), nanofluids can be assumed to be single phase fluids. However, almost all of new academic paper use two-phase assumption. Classical theory of single phase fluids can be applied, where physical properties of nanofluid is taken as a function of properties of both constituents and their concentrations. An alternative approach simulates nanofluids using a two-component model.

The spreading of a nanofluid droplet is enhanced by the solid-like ordering structure of nanoparticles assembled near the contact line by diffusion, which gives rise to a structural disjoining pressure in the vicinity of the contact line. However, such enhancement is not observed for small droplets with diameter of nanometer scale, because the wetting time scale is much smaller than the diffusion time scale.

Synthesis

Nanofluids are produced by several techniques they are, 1.Direct Evaporation (1 step), 2.Gas condensation/dispersion (2 step), 3.Chemical vapour condensation (1 step), 4.Chemical precipitation (1 step). Several liquids including water, ethylene glycol, and oils have been used as base fluids. Although stabilization can be a challenge, on-going research indicates that it is possible. Nano-materials used so far in nanofluid synthesis include metallic particles, oxide particles, carbon nano-tubes, graphene nano-flakes and ceramic particles.

Smart Cooling Nanofluids

Realizing the modest thermal conductivity enhancement in conventional nanofluids, a team of researchers at Indira Gandhi Centre for Atomic Research Centre, Kalpakkam developed a new class of magnetically polarizable nanofluids where the thermal conductivity enhancement up to 300% of basefluids is demonstrated. Fatty-acid-capped magnetite nanoparticles of different sizes (3-10 nm) have been synthesized for this purpose. It has been shown that both the thermal and rheological properties of such magnetic nanofluids are tunable by varying the magnetic field strength and orientation with respect to the direction of heat flow. Such response stimuli fluids are reversibly switchable and have applications in miniature devices such as micro- and nano-electromechanical systems. In 2013, Azizian et al. considered the effect of an external magnetic field on the convective heat transfer coefficient of water-based magnetite nanofluid experimentally under laminar flow regime. Up to 300% enhancement obtained at Re=745 and magnetic field gradient of 32.5 mT/mm. The effect of the magnetic field on the pressure drop was not as significant.

Nanoparticle Migration

In nanofluids, it is recognized that nanoparticles do not follow the fluid streamlines passively. In fact, there are some reasons that induce a slip velocity between the nanoparticles and the base fluid. Movements of nanoparticles has significant impact on rheological and thermophysical properties of the nanofluids. Therefore, investigating the nanoparticles motion is critical for evaluating the performance of nanoparticles inclusion to the base fluid as a heat transfer medium. Since the nanoparticles are very small (100 nm), Brownian and thermophoretic diffusivities are the main slip mechanisms in nanofluids, as Buongiorno declared. Brownian diffusion is due to random drifting of suspended nanoparticles in the base fluid which originates from continuous collisions among the nanoparticles and liquid molecules. Thermophoresis induces nanoparticle migration from warmer to colder region (in opposite direction of the temperature gradient), making a non-uniform nanoparticle volume fraction distribution.

In fact, theoretical models estimated that nanoparticles are non-homogeneously distributed. The level of non-uniformity is completely depend on thermal boundary conditions, the nanoparticle size, shape, and material. Rigorous readers encouraged to find more interesting results in open literature.

Response Stimuli Nanofluids for Sensing Applications

Researchers have invented a nanofluid-based ultrasensitive optical sensor that changes its colour on exposure to extremely low concentrations of toxic cations. The sensor is useful in detecting minute traces of cations in industrial and environmental samples. Existing techniques for monitoring cations levels in industrial and environmental samples are expensive, complex and time-consuming. The sensor is designed with a magnetic nanofluid that consists of nano-droplets with magnetic grains suspended in water. At a fixed magnetic field, a light source illuminates the nanofluid where the colour of the nanofluid changes depending on the cation concentration. This color change occurs within a second after exposure to cations, much faster than other existing cation sensing methods.

Such response stimulus nanofluids are also used to detect and image defects in ferromagnetic components. The photonic eye, as it has been called, is based on a magnetically polarizable nano-emulsion that changes colour when it comes into contact with a defective region in a sample. The device might be used to monitor structures such as rail tracks and pipelines.

Magnetically Responsive Photonic Crystals Nanofluids

Magnetic nanoparticle clusters or magnetic nanobeads with the size 80–150 nanometers form ordered structures along the direction of the external magnetic field with a regular interparticle spacing on the order of hundreds of nanometers resulting in strong diffraction of visible light in suspension.

Applications

Nanofluids are primarily used for their enhanced thermal properties as coolants in heat transfer equipment such as heat exchangers, electronic cooling system(such as flat plate) and radiators. Heat transfer over flat plate has been analyzed by many researchers. However, they are also useful for their controlled optical properties. Graphene based nanofluid has been found to enhance Polymerase chain reaction efficiency. Nanofluids in solar collectors is another application where nanofluids are employed for their tunable optical properties.

Bifunctional Plasmonic-magnetic Nanoparticles

Bifunctional plasmonic-magnetic nanoparticles (PM-NPs) consist of optical and magnetic components. They are useful for biomedical applications that require optical sensing/imaging/heating, magnetic stimulation/manipulation, or both of these functionalities. An example of these functionalities is that PM-NPs can attach to biological entities and separate them under an external magnetic field while simultaneously detecting their chemical nature via optical sensing. These functionalities are especially useful when studying tissues deep within an organism. Current available bifunctional PM-NPs are limited to quasi-zero-dimensional (0D) nanostructures. There are few reports of quasi-one-dimensional (1D) nanostructures, although 1D PM nanotubes provide advantages for biomedical applications that are not available with 0D nanostructures. A specific type of 1D PM nanotube has applications in targeted single-cell sensing.

SERS Characterization

A laser beam focused on two neighboring Ag nanoparticles can generate a high intensity electric field in the narrow junction due to LSP resonance as a result of coherent electron oscillation in the Ag nanoparticles. This junction is referred to as a "hotspot". If a molecule is in the vicinity of the hotspot, it's Raman scattering signals can be significantly amplified. This phenomenon is called surface enhanced Raman scattering (SERS). SERS EF is determined by the electric field intensity. Nanocapsules that have Ag nanoparticles on the inside and outside (double-layer nanotubes) exhibit 2x SERS intensity than those with a single layer of Ag nanoparticles.

Manufacturing SERS Nanocapsules and Arrays

Researchers at the University of Texas at Austin found a method for economically synthesizing SERS nanocapsules and creating arrays of the nanocapsules capable of ultrasensitive and position-predictable SERS detections. In doing so, the researchers also tackled two major problems previously associated with SERS sensing: 1. Lack of many hotspots with controlled gaps 2. Difficulty of assembling SERS probes at designated locations.

The nanocapsules created in this study comprised a 3-layer structure: the Ag/Ni/Ag nano core, a silica capsule, a uniformly distributed Ag nanoparticles on silica. Creating the nanocapsule works as follows: (1) The electrodeposition of a nanoporous anodized aluminum oxide and subsequent deposition of a Cu layer onto this membrane to create a three-electrode electrodeposition system; (2) building the nanowires via electrodeposition, starting at the bottom of the membrane's nanopores (the amount of electric charge passing through the circuit controls the length of the Ag/Ni/Ag nanowires; (3) dissolving the template membrane in NaOH and subsequently washing the nanowires via sonication and centrifuging in ethanol and deionized water; (4) coating nanocore in silica via hydrolysis of tetraethyl orthosilicate; (5) isolating Ag ions and then incubating with the silica-coated nanocapsule to reduce ionic Ag into metallic Ag nanoparticles on the silica.

In terms of functionality, the nanocore of Ag and Ni allows for manipulating the nanocapsule with electric tweezers. Ag/Ni/Ag nanocores prove necessary for steering/orientation via the AC field. The silica capsule serves to separate the nanocore from the Ag nanoparticles on the outside, preventing the plasmonic quenching effect. The Ag nanodots act as a hotspot layer (with an enhanced electric field) and perform the sensing task.

The researchers arranged the nanocapsules into arrays using magnets of Cr/Ni/Ag. The Cr anchored the magnet in substrate; the Ni worked in concert with the Ni in the nanocapsule core by providing an attractive magnetic field; the Au layer tuned the magnetic interaction force between nanomagnet and nanocapsule. The researchers used a quadruple electrode chip to create an array of nanocapsules. The AC and DC fields could be oriented parallel or perpendicular to each other in order to create the electric tweezers (the mechanism could transport capsules parallel or perpendicular to their orientations depending on whether or not AC and DC fields were directed parallel or perpendicular to one another). The Ag/Ni/Ag nanocores proved necessary for steering/orientation via the AC field (when removed no net torque resulted). The electric tweezers then moved the nanocapsules to their nanomagnet anchors, utilizing the attraction between Ni in capsule and magnet.

These nanocapsule arrays prove consistent, with the variation of Raman intensity between sensors +/-9%. This low error margin comes from the controlled sizing and spacing of Ag nanoparticles and the large number of such hotspots on the nanocapsule. Previously there existed only randomly placed hotspots, which produced more variation between nanocapsule sensitivity and therefore increased the error margin. The nanocapsules can detect Raman spectra at small concentrations (for example, 1,2-bi-(4-pyridyl) ethylene (BPE) at 10-12 M) and as concentration increases, the detection sensitivity increases logarithmically.

Plasmonic Simulation

SERS enhancement can be attributed to two factors: electric field (E-field) enhancement due to the plasmonic resonance of nanoparticles and chemical enhancement due to charge transfer between the molecules and metal particles. E-field enhancement is a major contributor to SERS. A group of metallic particles on the same surface can provide stronger E-field enhancement than that of single particles or dimers. A group of nanoparticles with controlled distributions further enhance the E field. Narrower junctions between nanoparticles exhibit a strong E field. For smaller nanoparticles, the normalized electric field increases as size increases. However, for nanoparticles of diameters >40 nm, E-field enhancement reduces. This is because the plasmonic resonance shifts to a longer wavelength. When a double-layer nanotube is placed on top of a glass substrate and excited by a surface normal beam, the hot-spot is at the bottom of the nanotube. This is because the bottom hot-spot is surrounded by silica, while the top is surrounded by air. In addition, the inter-particle coupling through the nanoparticle chain at the outer surface enhances the electric field at the bottom of the nanotube. Not only does the inner layer add more hot-spots for SERS sensing, but it also increases the intensity of the hot spots in the outer layer nanoparticles. However, when the core is filled with platinum, the opposite is observed. The hot spots at the top of the nanotube are enhanced by 2x, while those at the bottom are reduced. This is because at the top of the nanotube, light is reflected to enhance the hot spot, while no light is able to pass through the platinum, which reduces the hot spots on the bottom. A similar phenomenon occurs when the tube is filled with nickel. Thickness of the silica also affects enhancement of SERS. Near-field enhancement can be observed on nanotubes with a silica coating ranging from 70 to 150 nm. However, if the thickness is increased, near-field enhancement will disappear.

Magnetic Characterization

These 1D PM nanotubes also offer tunable magnetic properties for controlled manipulation. When the Ni segment is relatively large, the anisotropy direction is along the long axis. When the Ni segment is relatively low, magnetic anisotropy is transverse to the length of the nanotubes with zero remanence. This is a way of controlling the magnetic anisotropy, and can be used to affect the way the tubes align with a magnetic field.

Single-Cell Bioanalysis

Because of their various functionalities, 1D PM nanotubes can be used for various applications, including analysis of membrane composition of a single animal cell among many. Because of the bifunctionality of the nanotubes, it is possible to precisely transport the nanotube to a specific living cell, and analyze various features of that cell.

Nanoshell

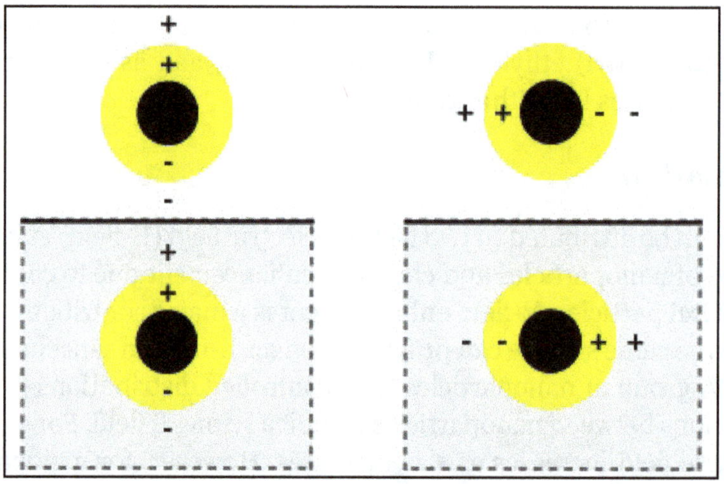

A nanoshell, or rather a nanoshell plasmon, is a type of spherical nanoparticle consisting of a dielectric core which is covered by a thin metallic shell (usually gold). These nanoshells involve a quasiparticle called a plasmon which is a collective excitation or quantum plasma oscillation where the electrons simultaneously oscillate with respect to all the ions.

The simultaneous oscillation can be called plasmon hybridization where the tunability of the oscillation is associated with mixture of the inner and outer shell where they hybridize to give a lower energy or higher energy. This lower energy couples strongly to incident light, whereas the higher energy is an anti-bonding and weakly combines to incident light. The hybridization interaction is stronger for thinner shell layers, hence, the thickness of the shell and overall particle radius determines which wavelength of light it couples with. Nanoshells can be varied across a broad range of the light spectrum that spans the visible and near infrared regions. The interaction of light and nanoparticles affects the placement of charges which affects the coupling strength. Incident light polarized parallel to the substrate gives a s-polarization, hence the charges are further from the substrate surface which gives a stronger interaction between the shell and core. Otherwise, a p-polarization is formed which gives a more strongly shifted plasmon energy causing a weaker interaction and coupling.

Discovery

The discovery of the nanoshell was made by Professor Naomi J. Halas and her team at Rice University in 2003. When she and her team discovered nanoshells, they weren't initially sure what potential such nanoshells held. "We said, 'Gee, what could it be good for?'" Halas told CNN. After many suggestions, cancer therapy came out of ongoing collaborations with bioengineers looking for different types of biomedical applications. "One of our visions", Halas stated, "no less than single visit diagnosis and treatment of cancer". In 2003 Halas was awarded for Best Discovery of 2003 by *Nanotechnology Now*.

Production

A state of the art method for synthesizing gold nanoshells is the use of the Microfluidic Composite Foams. This method has the potential to replace the standard lithographic method of synthesizing plasmonic nanoshells. The production process described below was an experiment performed by Suhanya Duraiswamy and Saif A. Khan of the Department of Chemical and Biomolecular Engineering in Singapore. Although this method was an experiment, it represents the future of nanoshells synthesis.

The materials required for the production of the nanoshells are the following; Tetraethyl orthosilicate, ammonium hydroxide, hydroxylamine hydrochloride, 3-aminopropyl tris, hydrogentetrachloroaurate(III) trihydrate, tetrakis(hydroxymethyl) phosphonium chloride, sodium hydroxide, potassium carbonate, ethanol, Ultrapure water and glassware washed in aqua regia and rinsed thoroughly in water.

The first step in synthesizing nanoshells in this method is by creating the device for the reaction to take place within. Microfluidic device patterns were fabricated onto silicon wafers by standard photolithography using negative photoresist SU-8 2050. Devices were subsequently molded in poly(dimethyl siloxane) (PDMS) using the soft lithography technique.(40) Briefly, PDMS was molded onto the SU-8 masters at 70 °C for 4 h, peeled, cut, and cleaned. Inlet and outlet holes (1/16-in. o.d.) were punched into the device. The microchannels were irreversibly bonded to a glass slide precoated with a thin layer of PDMS after a brief 35 s air plasma treatment. The microchannels have rectangular cross-section and are 300 μm wide, 155 μm deep, and 0.45 m long.

The actual production of the nanoparticles involves pumping "silicone oil, a mixture of gold-seeded silica particles and gold-plating solution and reducing agent solution to the microfluidic device while nitrogen gas was delivered from a cylinder." The plating solution was then left to age, in a controlled environment, for longer than 24 hours. After the aging process, the fluid is collected from the Microfluidic Device and placed in a centrifuge. The resulting liquid has a layer of oil on the surface with a solution below that contains the nanoshells.

The reason this method is revolutionary is that the size and relative thickness of the gold nanoshell can be controlled by changing the amount of time the reaction is allowed to take place as well as the concentration of the plating solution. Thereby allowing researchers to tailor the particles to suit their given needs. Albeit for optics or cancer treatment.

Cancer Treatment

Gold-shelled nanoparticles, which are spherical nanoparticles with silica cores and gold shells, are used in cancer therapy and bio-imaging enhancement. Theranostic probes – capable of detection and treatment of cancer in a single treatment - are nanoparticles that have binding sites on their shell that allow them to attach to a desired location (typically cancerous cells) then can be imaged through dual modality imagery (an imaging strategy that uses x-rays and radionuclide imaging) and through near-infrared fluorescence. The reason gold nanoparticles are used is due to their vivid optical properties which are controlled by their size, geometry, and their surface plasmons. Gold nanoparticles (such as AuNPs) have the benefit of being biocompatible and the flexibility to have multiple different molecules, and fundamental materials, attached to their shell (almost anything that can normally be attached to gold can be attached to the gold nano-shell, which can be used in helping identifying and treating cancer). The treatment of cancer is possible only because of the scattering and absorption that occurs for plasmonics. Under scattering, the gold-plated nano-particles become visible to imaging processes that are tuned to the correct wavelength which is dependent upon the size and geometry of the particles. Under absorption, photothermal ablation occurs, which heats the nanoparticles and their immediate surroundings to temperatures capable of killing the cancer cells. This is accomplished with minimal damage to cells in the body due to the utilization of the "water window" (the spectral range between 800 and 1300 nm). As the human body is mostly water, this optimizes the light used versus the effects rendered.

These gold nanoshells are shuttled into tumors by the use of phagocytosis, where phagocytes engulf the nanoshells through the cell membrane to form an internal phagosome, or macrophage. After this it is shuttled into a cell and enzymes are usually used to metabolize it and shuttle it back out of the cell. These nanoshells are not metabolized so for them to be effective they just need to be within the tumor cells and photo-induced cell death (as described above) is used to terminate the tumor cells.

Nanoparticle-based therapeutics have been successfully delivered into tumors by exploiting the enhanced permeability and retention effect, a property that permits nanoscale structures to be taken up passively into tumors without the assistance of antibodies. Delivery of nanoshells into the important regions of tumors can be very difficult. This is where most nanoshells try to exploit the tumor's natural recruitment of monocytes for delivery as seen in the figure. This delivery system is called a "Trojan Horse".

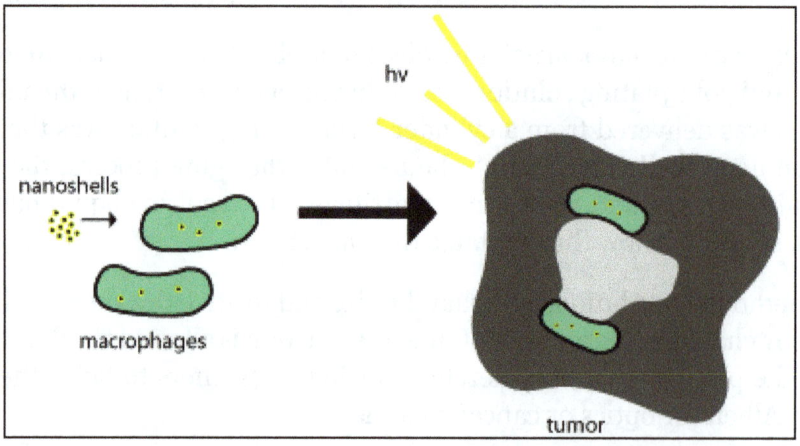

This process works so well since tumors are about 3/4 macrophages and once monocytes are brought into the tumor, it differentiates into macrophages which would also be need to maintain the cargo nanoparticles. Once the nanoshells are at the necrotic center, near-infrared illumination is used to destroy the tumor associated macrophages.

Additionally, these nanoparticles can be made to release antisense DNA oligonucleotides when under photo-activation. These oligonucleotides are used in conjunction with the photo-thermal ablation treatments to perform gene-therapy. This is accomplished because nanoparticle complexes are delivered inside of cells then undergo light induced release of DNA from their surface. This will allow for the internal manipulation of a cell and provide a means for monitoring a group cells return to equilibrium.

Another example of nanoshell plasmonics in cancer treatment involves placing drugs inside of the nanoparticle and using it as a vehicle to deliver toxic drugs to cancerous sites only. This is accomplished by coating the outside of a nanoparticle with iron oxide (allowing for easy tracking with an MRI machine), then once the area of the tumor is coated with the drug-filled nanoparticles, the nanoparticles can be activated using resonant light waves to release the drug.

Nanosized Aluminium Oxide

Nanosized aluminium oxide (nanosized alumina) occurs in the form of spherical or nearly spherical nanoparticles, and in the form of oriented or undirected fibers.

Properties

Properties, of the final material, defined as the set of properties of the solid Aluminium oxide and specific properties of nanostructures.

Properties of Nanoscale Colloidal Alumina Particles

- Small diameter of the particles/fibers (2-10 nm).
- High specific surface area (>100 m^2/g).
- High defectiveness of the material surface and specific structure of the nanoparticles (the volume and size of pores, degree of crystallinity, phase composition, structure, and composition of the surface — modification possibility).

Properties of the Nanoscale Fibers of Aluminium Oxide

- The ratio of length-diameter about 20,000,000:1.
- A high degree of orientation of the fibers.
- The weak interaction of the fibers among themselves.
- Absence of surface pores.
- High surface concentration of hydroxyl groups.

Production

Methods of obtaining Powders of Aluminium Oxide Nanometer Scale

Industrially obtained alumina nano fibers of Nafen brand.

1. Grinding powder alumina particles of a nanometer level (for example, 10-50 nm). For example, using a planetary mill using grinding bodies of size less than 0.1 μm.

2. The decomposition of fresh chemically-synthesized AlOOH or $Al(OH)_3$ to aluminium oxide in the rapid achievement of the temperature of decomposition 175 °C and use for it the pressure of 5 bars within thirty minutes. The sooner of the temperature of decomposition of the hydroxo-compounds of aluminium is achieved, the smaller the resulting particles nano oxide in size.

Alumina Nano Fibers

The oxidation of the surface of some liquid metal alloys leads to the formation of loose or porous 3D nanostructures. For the first time this effect was observed in the system aluminium-mercury and published more than 100 years ago. The fibers of such kind do not occur in nature and only grown by artificial means. Depending on the method of synthesis can be produced various nanostructures, such as aerogel from oxyhydroxide aluminium (AlOOH or Al_2O_3*nH_2O, where n=1-4, are easily turned into aluminium oxide) or nano-fibers of aluminium oxide (Al_2O_3).

At this moment, the main ways of production are:

1. The method for selective oxidation of aluminium on the surface of the molten Ga-AI in a humid atmosphere at a temperature of 20 to 70 °C (Method of IPCE RAS).

2. The liquid metal technology of synthesis of nanostructural aerogel AlOOH from molten Ga-Bi and Al-Al (Institute of RF IPPE named after A. I. Leipunsky, Obninsk city).

3. Growing fiber nano oxide of aluminium on the surface of the aluminum melt (a Method of industrial synthesis, developed and patented by the ANF Technology).

Application

- Adsorbent (to capture hydrocarbon impurities from the air; for extracting fluorine from a variety of media (the ability of aluminum oxide to chemosensitivity fluorine ions used for the purification of water with increased fluorine content; for vapor recovery of hydrogen fluoride from gases of super phosphate and electrolysis) for brightening solutions in sugar production; to trap solvents; adsorption purification of oils (first transformer); adsorbent for gas and liquid adsorption chromatography (adsorption); for ion-exchange and sediment-sorption chromatography in aqueous solution (ion exchange and precipitation); as an inert carrier during liquid-distribution chromatography)
- Desiccant (for drying of gases (deep dehydration to dew points of -60 °C and below); the preservation of instruments and equipment, and also for such systems as the respiratory valves, tanks, transformers, etc.; to create protective atmospheres during long-term storage of food and pharmaceutical products)
- Sorbent of the ions of metals from solutions of their salts, for example, $CsNO_3$, $AgNO_3$, $Ba(NO_3)_2$, $Sr(NO_3)_2$, $Pb(NO_3)_2$, etc., with the possibility of obtaining of metal oxides on the surface of the fibers during annealing
- Sorbent of radionuclides from wastewaters of nuclear power plants
- Inert (reinforcing) filler
- Ceramics and composites (including composite metals) — high toughness, fire resistance and anti-friction properties, insulating properties. Known use in several products such as burner discharge lamp, the substrate of integrated circuits, shut-off elements ceramic pipeline valves, prostheses, etc.
- Abrasive (composed of means for ultra-fine polishing)
- Refractory (high-temperature component for heat insulation)

In addition to these areas, used as a catalyst and carrier of catalysts. Nanoscale oxide due to the small diameter of the particles/fibers, high specific surface area and activity associated with the defects, and the specific structure of the nanoparticles (the volume and size of pores, degree of crystallinity, phase composition, structure, and composition of the surface) strongly enhances the catalytic properties, and increases the range of massive aluminium oxide as a catalyst.

Nanoparticle–biomolecule Conjugate

A nanoparticle–biomolecule conjugate is a nanoparticle with biomolecules attached to its surface. Nanoparticles are minuscule particles, typically measured in nanometers (nm), that are used in nanobiotechnology to explore the functions of biomolecules. Properties of the ultrafine particles are characterized by the components on their surfaces more so than larger structures, such as cells, due to large surface area-to-volume ratios. Large surface area-to-volume-ratios of nanoparticles optimize the potential for interactions with biomolecules.

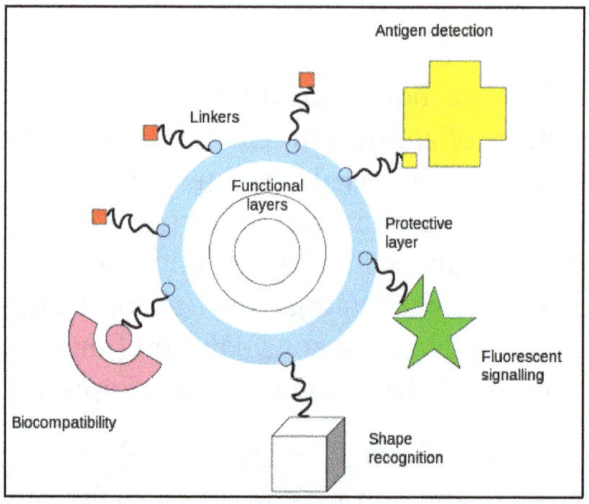

Attachments on nanoparticles make them more biocompatible.

Characterization

Major characteristics of nanoparticles include volume, structure, and visual properties that make them valuable in nanobiotechnology. Depending on specific properties of size, structure, and luminescence, nanoparticles can be used for different applications. Imaging techniques are used to identify such properties and give more information about the tested sample. Techniques used to characterize nanoparticles are also useful in studying how nanoparticles interact with biomolecules, such as amino acids or DNA, and include magnetic resonance imaging (MRI), denoted by the solubility of the nanoparticles in water and fluorescent. MRI can be applied in the medical field to visualize structures; atomic force microscopy (AFM) that gives a topographic view of the sample on a substrate; transmission electron microscopy (TEM) that gives a top view, but with a different technique then that of atomic force microscopy; Raman spectroscopy or surface enhanced Raman spectroscopy (SERS) gives information about wavelengths and energy in the sample. ultraviolet-visible spectroscopy (UV-Vis) measures the wavelengths where light is absorbed; X-Ray diffraction (XRD) generally gives an idea of the chemical composition of the sample.

Chemistry

Physical

Nanomolecules can be created from virtually any element, but the majority produced in today's industry use carbon as the basis upon which the molecules are built around. Carbon can bond with nearly any element, allowing many possibilities when it comes to creating a specific molecule. Scientists can create thousands upon thousands of individual nanomolecules from a simple carbon basis. Some of the most famous nanomolecules currently in existence are solely carbon; these include carbon nanotubes and buckminsterfullerenes. In contrast with nanomolecules, the chemical components of nanoparticles usually consist of metals, such as iron, gold, silver, and platinum.

Interactions between nanoparticles and molecules change depending on the nanoparticle's core. Nanoparticle properties depend not only on the composition of the core material, but also on varying thicknesses of material used. Magnetic properties are particularly useful in molecule manipulation, and thus metals are often used as core material. Metals contain inherent magnetic properties that allow for manipulation of molecular assembly. As nanoparticles interact with molecules via ligand properties, molecular assembly can be controlled by external magnetic fields interacting with magnetic properties in the nanoparticles. Significant problems with producing nanoparticles initially arise once these nanoparticles are generated in solution. Without the use of a stabilizing agent, nanoparticles tend to stick together once the stirring is stopped. In order to counteract this, a certain collidial stabilizer is generally added. These stabilizers bind to the nanoparticles in a way that prevents other particles from bonding with them. Some effective stabilizers found so far include citrate, cellulose, and sodium borohydride.

Application Chemistry

Nanoparticles are desirable in today's industry for their high surface area-to-volume ratio in comparison with larger particles of the same elements. Because chemical reactions occur at a rate directly proportional to the available surface area of reactant compounds, nanoparticles can generate reactions at a much faster rate than larger particles of equal mass. Nanoparticles therefore are among the most efficient means of producing reactions and are inherently valuable in the chemical industry. The same property makes them valuable in interactions with molecules.

Applications with Biomolecules and Biological Processes

Nanoparticles have the potential to greatly influence biological processes. The potency of a nanoparticle increases as its surface area-to-volume-ratio does. Attachments of ligands to the surface of nanoparticles allow them to interact with biomolecules.

Identification of Biomolecules

Nanoparticles are valuable tools in identification of biomolecules, through the use of bio-tagging or labeling. Attachments of ligands or molecular coatings to the surface of a nanoparticle facilitate nanoparticle-molecule interaction, and make them biocompatible. Conjugation can be achieved through intermolecular attractions between the nanoparticle and biomolecule such as covalent bonding, chemisorption, and noncovalent interactions.

To enhance visualization, nanoparticles can also be made to fluoresce by controlling the size and shape of a nanoparticle probe. Fluorescence increases luminescence by increasing the range of wavelengths the emitted light can reach, allowing for biomarkers with a variety of colors. This technique is used to track the efficacy of protein transfer both in vivo and in vitro in terms of genetic alternations.

Biological Process Control

Biological processes can be controlled through transcription regulation, gene regulation, and enzyme inhibition processes that can be regulated using nanoparticles. Nanoparticles can play a part in gene regulation through ionic bonding between positively charged cationic ligands on the surfaces of nanoparticles and negatively charged anionic nucleic acids present in DNA. In an experiment, a nanoparticle-DNA complex inhibited transcription by T7 RNA polymerase, signifying strong bonding in the complex. A high affinity of the nanoparticle-DNA complex indicates strong bonding and a favorable use of nanoparticles. Attaching ionic ligands to nanoparticles allows control over enzyme activity. An example of enzyme inhibition is given by binding of a-chymotrypsin (ChT), an enzyme with a largely cationic active site. When a-chymotrypsin is incubated with anionic (negatively charged) nanoparticles, ChT activity is inhibited as anionic nanoparticles bind to the active site. Enzyme activity can be restored by the addition of cationic surfactants. Alkyl surfactants form a bilayer around ChT, whereas thiol and alcohol surfactants alter the surface of ChT such that interactions with nanoparticles are interrupted. Though formation of a protein-nanoparticle complex can inhibit enzyme activity, studies show that it can also stabilize protein structure, and significantly protect the protein from denaturization. Attachments of ligands to segments of nanoparticles selected for functionalization of metallic properties can be used to generate a magnetic nanowire, which generates a magnetic field that allows for the manipulation of cellular assemblies.

Genetic Alteration

Nanoparticles can also be used in conjunction with DNA to perform genetic alterations. These are frequently monitored through the use of fluorescent materials, allowing scientists to judge if these tagged proteins have successfully been transmitted—for example green fluorescent protein, or GFP. Nanoparticles are significantly less cytotoxic than currently used organic methods, providing a more efficient method of monitoring genetic alternations. They also do not degrade or bleach with time, as organic dyes do. Suspensions of nanoparticles with the same size and shapes (mono-dispersed) with functional groups attached to their surfaces can also electrostatically bind to DNA, protecting them from several types of degradation. Because the fluorescence of these nanoparticles does not degrade, cellular localization can be tracked without the use of additional tagging, with GFPs or other methods. The 'unpacking' of the DNA can be detected in live cells using luminescence resonance energy transfer (LRET) technology.

Medical Implications

Small molecules *in vivo* have a short retention time, but the use of larger nanoparticles does not. These nanoparticles can be used to avoid immune response, which aids in the treatment of chronic diseases. It has been investigated as a potential cancer therapy and also has the potential to affect the understanding of genetic disorders. Nanoparticles also have the potential to aid in site-specific drug delivery by improving the amount of unmodified drug that is circulated within the system, which also decreases the necessary dosage frequency. The targeted nature of nanoparticles also means that non-targeted organs are much less likely to experience side effects of drugs intended for other areas.

Another example is the delivery of nucleic acids to alter key genes associated with cancer progression. For example, Conde et al have shown that by twisting RNA strands into a triple helix and embedding them in a biocompatible gel, they can not only deliver the strands efficiently but also use them to shrink aggressive tumors in mice. Using this technique, the researchers dramatically improved cancer survival rates by simultaneously turning on a tumor-suppressing microRNA and de-activating one that causes cancer. They believe their approach could also be used for delivering other types of RNA, as well as DNA and other therapeutic molecules.

Studying Cell Interactions

Cellular interactions occur at a microscopic level and cannot be easily observed even with the advanced microscopes available today. Due to difficulties observing reactions at the molecular level, indirect methods are used which greatly limits the scope of the understanding that can be gained by studying these processes essential to life. Advances in the material industry has evolved a new field known as nanobiotechnology, that uses nanoparticles to study interactions at the biomolecular level.

One area of research featuring nanobiotechnology is the extracellular matrices of cells (ECM). The ECM is primarily composed of interwoven fibers of collagen and elastin that have diameters ranging from 10–300 nm. In addition to holding the cell in place, the ECM has a variety of other functions including providing a point of attachment for the ECM of other cells and transmembrane receptors that are essential for life. Until recently it has been nearly impossible to study the physical forces that help cells maintain their functionality, but nanobiotechnology has given us the ability to learn more about these interactions. Using the unique properties of nanoparticles, it is possible to control how the nanoparticles adhere to certain patterns present in the ECM, and as a result can understand how changes in the ECM's shape can affect cell functionality.

Using nanobiotechnology to study the ECM allows scientists to investigate the binding interactions that occur between the ECM and its supporting environment. Investigators were able to study these interactions by utilizing tools such as optical tweezers, which have the ability to trap nano-scale objects with focused light. The tweezers can affect the binding of a substrate to the ECM by attempting to draw the substrate away from it. The light emitted from the tweezers was used to restrain ECM-coated microbeads, and the changes in the force exerted by the ECM onto the substrate were studied by modulating the effect of the optical tweezers. Experiments showed that the force exerted by the ECM on the substrate positively correlated with the force of the tweezers, which led to the subsequent discovery that the ECM and the transmembrane proteins are able to sense external forces, and can adapt to overcome these forces.

Nanotechnology Crossing the Blood-brain Barrier

The blood-brain barrier (BBB) is composed of a system of capillaries that has an especially dense lining of endothelial cells which protects the central nervous system (CNS) against the diffusion of foreign substances into the cerebrospinal fluid. These harmful objects include microscopic bacteria, large hydrophobic molecules, certain hormones and neurotransmitters, and low-lipid-soluble molecules. The BBB prevents these harmful particles from entering the brain via tight junctions between endothelial cells and metabolic barriers. The thoroughness with which the BBB does its job makes it difficult to treat diseases of the brain such as cancer, Alzheimer's, and autism, because it is very difficult to transport drugs across the BBB. Currently, in order to deliver therapeutic molecules into the brain, doctors must use highly invasive techniques such as drilling directly into the brain, or sabotaging the integrity of the BBB through biochemical means. Due to their small size and large surface area, nanoparticles offer a promising solution for neurotherapeutics.

Nanotechnology is helpful in delivering drugs and other molecules across the blood-brain barrier (BBB). Nanoparticles allow drugs, or other foreign molecules, to efficiently cross the BBB by camouflaging themselves and tricking the brain into providing them with the ability to cross the BBB in a process called the Trojan Horse Method. Using nanotechnology is advantageous because only the engineered complex is necessary whereas in ordinary applications the active compound must carry out the reaction. This allows for maximum efficacy of the active drug. Also, the use of nanoparticles results in the attraction of proteins to the surfaces of cells, giving cell membranes a biological identity. They also use endogenous active transport where transferrin, an iron binding protein, is linked to rod-shaped semiconductor nanocrystals, in order to move across the BBB into the brain. This discovery is a promising development towards designing an efficient nanoparticle-based drug delivery system.

Self-assembly of Nanoparticles

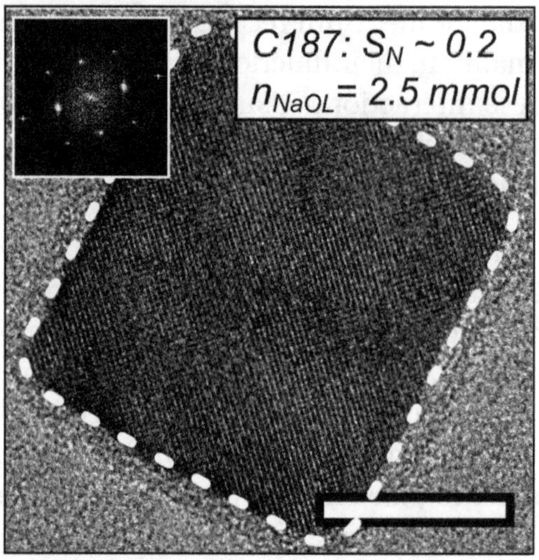

Transmission electron microscopy image of an iron oxide nanoparticle. Regularly arranged dots within the dashed border are columns of Fe atoms. Left inset is the corresponding electron diffraction pattern. Scale bar: 10 nm.

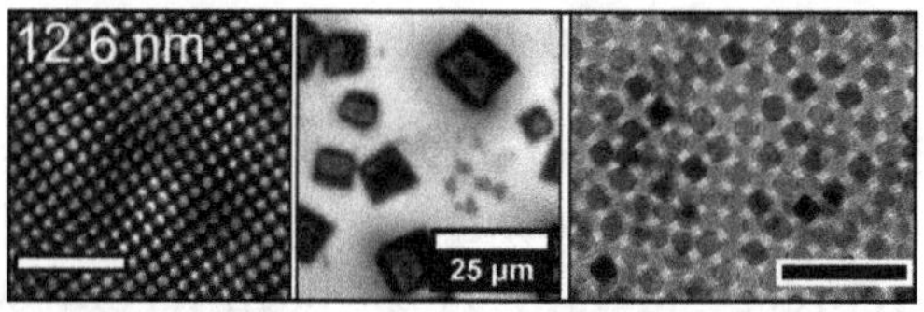

Iron oxide nanoparticles can be dispersed in an organic solvent (toluene). Upon its evaporation, they may self-assemble (left and right panels) into micron-sized mesocrystals (center) or multilayers (right). Each dot in the left image is a traditional "atomic" crystal shown in the image above. Scale bars: 100 nm (left), 25 μm (center), 50 nm (right).

Self-assembly is a phenomenon where the components of a system assemble themselves spontaneously via an interaction to form a larger functional unit. This spontaneous organization can be due to direct specific interaction and/or indirectly through their environment. Due to the increasing technological advancements, the study of materials in the nanometre scale is becoming more important. The spatial arrangements of these self-assembled nanoparticles can be potentially used to build increasingly complex structures leading to a wide variety of materials that can be used for different purposes.

At the molecular level, intermolecular force hold the spontaneous gathering of molecules into a well-defined and stable structure together. In chemical solutions, self-assembly is an outcome of random motion of molecules and the affinity of their binding sites for one another. In the area of nanotechnology, developing a simple, efficient method to organize molecules and molecular clusters into precise, pre-determined structure is crucial.

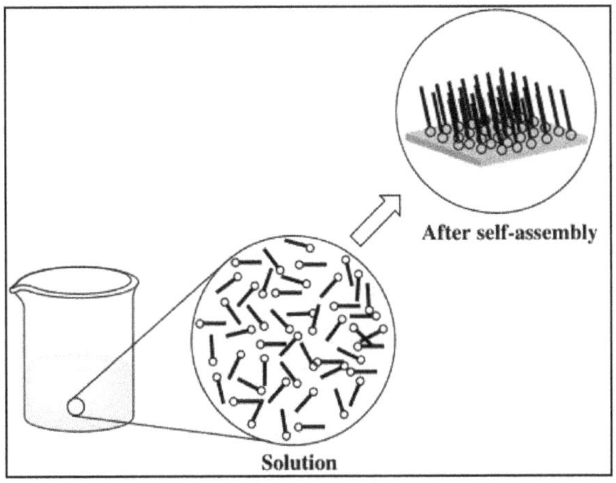

An example of self-assembly of nanoparticles in a solution. In this diagram, it can be seen that a disordered system formed an organized structure which can be due to specific interactions among the particles.

History

The study of self-assembly of nanoparticles is a new area in nanotechnology research. Scientists who are focusing their study on nanotechnology recognized that some properties of atoms and molecules enable them to arrange themselves into patterns. There are a variety of applications where the self-assembly of nanoparticles can be useful. For example, building sensors to detect chemical and biological molecules. In addition, it can also be used on creating computer chips with smaller component sizes, which can then allow more computing power to be stored on a chip. Recently, researchers were able to watch nanoparticles self-assemble for the very first time in real-time. This research was done by scientists at the U.S. Department of Energy's (DOE) Argonne National Laboratory. They used a transmission electron microscope (TEM) located at the Center for Nanoscale Materials to capture the quick movement of the nanoparticles into a self-assembled system.

Thermodynamics

Self-assembly is an equilibrium process where the assembled components are in equilibrium with the individual components. Self-assembly is driven by the minimization of Gibbs free energy. The minimization of Gibbs free energy is attained by the minimization of repulsive and the maximization of attractive molecular interactions. In addition, the lower free energy is usually a result of a weaker intermolecular force between self-assembled moieties and is essentially enthalpic in nature.

The thermodynamics of the self-assembly process can be represented by a simple Gibbs free energy equation:

$$\Delta G_{SA} = \Delta H_{SA} - T\Delta S_{SA}$$

where if ΔG_{SA} is negative, self-assembly is a spontaneous process. ΔH_{SA} is the enthalpy change of the process and is largely determined by the potential energy/intermolecular forces between the assembling entities. ΔS_{SA} is the change in entropy associated with the formation of the ordered arrangement. In general, the organization is accompanied by a decrease in entropy and in order for the assembly to be spontaneous the enthalpy term must be negative and in excess of the entropy term. This equation shows that as the value of $T\Delta S_{SA}$ approaches the value of ΔH_{SA} and above a critical temperature, the self-assembly process will become progressively less likely to occur and spontaneous self-assembly will not happen.

The self-assembly is governed by the normal processes of nucleation and growth. Small assemblies are formed because of their increased lifetime as the attractive interactions between the components lower the Gibbs free energy. As the assembly grows, the Gibbs free energy continues to decrease until the assembly becomes stable enough to last for a long period of time. The necessity of the self-assembly to be an equilibrium process is defined by the organization of the structure which requires non-ideal arrangements to be formed before the lowest energy configuration is found.

Defects

In a self-assembled structure a reasonable concentration of defects is present. Defects can be defined thermodynamically. In most cases, the thermodynamic driving force for self-assembly is provided by weak intermolecular interactions and is usually of the same order of magnitude as the entropy term. In order for a self-assembling system to reach the minimum free energy configuration, there has to be enough thermal energy to allow the mass transport of the self-assembling molecules. For defect formation, the free energy of single defect formation is given by:

$$\Delta G_{DF} = \Delta H_{DF} - T\Delta S_{DF}$$

The enthalpy term, ΔH_{DF} does not necessarily reflect the intermolecular forces between the molecules, it is the energy cost associated with disrupting the pattern and may be thought of as a region where optimum arrangement does not occur and the reduction of enthalpy associated with ideal self-assembly did not occur. An example of this can be seen in a system of hexagonally packed cylinders where defect regions of lamellar structure exist.

If ΔG_{DF} is negative, there will be a finite number of defects in the system and the concentration will be given by:

$$\frac{N}{N_0} = exp(\frac{-\Delta E_{act}}{RT})$$

N is the number of defects in a matrix of N_0 self-assembled particles or features and ΔE_{act} is the activation energy of defect formation. The activation energy, ΔE_{act}, should not be confused with ΔH_{DF}. The activation energy represents the energy difference between the initial ideally arranges state and a transition state towards the defective structure. At low defect concentrations, defect formation is entropy driven until a critical concentration of defects allows the activation energy term to compensate for entropy. There is usually an equilibrium defect density indicated at the minimum free energy. The activation energy for defect formation increases this equilibrium defect density.

Particle Interaction

Intermolecular forces govern the particle interaction in self-assembled systems. The forces tend to be intermolecular in type rather than ionic or covalent because ionic or covalent bonds will "lock" the assembly into non-equilibrium structures. The types intermolecular forces seen in self-assembly processes are van der Waals, hydrogen bonds, and weak polar forces, just to name a few. In self-assembly, regular structural arrangements are frequently observed, therefore there must be a balance of attractive and repulsive between molecules otherwise an equilibrium distance will not exist between the particles. The repulsive forces can be electron cloud-electron cloud overlap or electrostatic repulsion.

Processing

The processes by which nanoparticles self-assemble are widespread and important. Understanding why and how self-assembly occurs is key in reproducing and optimizing results. Typically, nanoparticles will self-assemble for one or both of two reasons: molecular interactions and external direction. Recently, researchers are more interested in the latter of these two reasons, and are able to apply external directing to a large variety of systems.

Self-assembly by Molecular Interactions

Nanoparticles have the ability to assemble chemically through covalent or noncovalent interactions with their capping ligand. The terminal functional group(s) on the particle are known as capping ligands. As these ligands tend to be complex and sophisticated, self-assembly can provide a simpler pathway for nanoparticle organization by synthesizing efficient functional groups. For instance, DNA oligomers have been a key ligand for nanoparticle building blocks to be self-assembling via sequence-based specific organization.However, to deliver precise and scalable (programmable) assembly for a desired structure, a careful positioning of ligand molecules onto the nanoparticle counterpart should be required at the building block (precursor) level, such as direction, geometry, morphology, affinity, etc.The successful design of ligand-building block units can play an essential role in manufacturing a wide-range of new nano systems, such as nanosensor systems, nanomachines/nanobots, nanocomputers, and many more uncharted systems.

Intermolecular Forces

Nanoparticles can self-assemble as a result of their intermolecular forces. As systems look to minimize their free energy, self-assembly is one option for the system to achieve its lowest free energy thermodynamically. Nanoparticles can be programmed to self-assemble by changing the functionality of their side groups, taking advantage of weak and specific intermolecular forces to spontaneously order the particles. These direct interparticle interactions can be typical intermolecular forces such as hydrogen bonding or Van der Waals forces, but can also be internal characteristics, such as hydrophobicity or hydrophilicity. For example, lipophilic nanoparticles have the tendency to self-assemble and form crystals as solvents are evaporated. While these aggregations are based on intermolecular forces, external factors such as temperature and pH also play a role in spontaneous self-assembly.

Hamaker Interaction

As nanoparticle interactions take place on a nanoscale, the particle interactions must be scaled similarly. Hamaker interactions take into account the polarization characteristics of a large number of nearby particles and the effects they have on each other. Hamaker interactions sum all of the forces between all particles and the solvent(s) involved in the system. While Hamaker theory generally describes a macroscopic system, the vast number of nanoparticles in a self-assembling system allows the term to be applicable. Hamaker constants for nanoparticles are calculated using Lifshitz theory, and can often be found in literature.

Hamaker Constants for Nanoparticles in Water	
Material	**A_{131}**
Fe_3O_4	22
Υ-Fe_2O_3	26
α-Fe_2O_3	29
Ag	33
Au	45
All values reported in zJ	

Externally Directed Self-assembly

The natural ability of nanoparticles to self-assemble can be replicated in systems that do not intrinsically self-assemble. Directed self-assembly (DSA) attempts to mimic the chemical properties of self-assembling systems, while simultaneously controlling the thermodynamic system to maximize self-assembly.

Electric and Magnetic Fields

External fields are the most common directors of self-assembly. Electric and magnetic fields allow induced interactions to align the particles. The fields take advantage of the polarizability of the nanoparticle and its functional groups. When these field-induced interactions overcome random Brownian motion, particles join to form chains and then assemble. At more modest field strengths, ordered crystal structures are established due to the induced dipole interactions. Electric and magnetic field direction requires a constant balance between thermal energy and interaction energies.

Flow Fields

Macroscopic Viscous Flow

Macroscopic viscous flow fields can direct self-assembly of a random solution of particles into ordered crystals. However, the assembled particles tend to disassemble when the flow is stopped or removed. Shear flows are useful for jammed suspensions or random close packing. As these systems begin in nonequilibrium, flow fields are useful in that they help the system relax towards ordered equilibrium. Flow fields are also useful when dealing with complex matrices that themselves have rheological behavior. Flow can induce anisotropic viseoelastic stresses, which helps to overcome the matrix and cause self-assembly.

Large Amplitude Oscillatory Shear (LAOS)

Large amplitude oscillatory shear (LAOS) is most effective for particles that are 100 nm-1 μm in size. Hard and soft shears can order in steady shear. However, this assembly strongly relies on particle volume fraction, particle interaction potentials, polydisterity, and shear rate and strain. The large amount of directing factors can cause complications in directing self-assembly by LAOS. Diblock copolymer micells have recently been studied in regard to structuring nanoparticles in bulk.

Combination of Fields

The most effective self-assembly director is a combination of fields. If the fields and conditions are optimized, self-assembly can be permanent and complete. When a field combination is used with nanoparticles that are tailored to be intrinsically responsive, the most complete assembly is observed. Combinations of fields allow the benefits of self-assembly, such as scalability and simplicity, to be maintained while being able to control orientation and structure formation. Field combinations possess the greatest potential for future directed self-assembly work.

Interfaces

Solid Interfaces

Nano-particles can self-assemble on solid surfaces after applying external forces (like magnetic, electric, and flow). Templates made of microstructures like carbon nanotubes or block polymers can also be used to assist in self-assembly; they cause directed self-assembly (DSA) in which active sites are embedded to selectively induce nanoparticle deposition. Such templates are considered as any object onto which different particles can be arranged into a structure with a morphology similar to that of the template. Carbon nanotubes (microstructures), single molecules, or block copolymers are common templates. Nanoparticles are often shown to self-assemble within distances of nanometers and micrometers, but block copolymer templates can be used to form well-defined self-assemblies over macroscopic distances. By incorporating active sites to the surfaces of nanotubes and polymers, the functionalization of these templates can be transformed to favor self-assembly of specified nanoparticles.

Liquid Interfaces

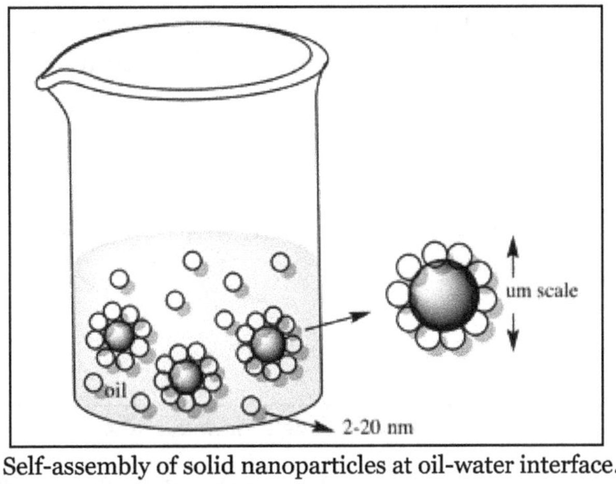

Self-assembly of solid nanoparticles at oil-water interface.

Pickering and Ramsden explained the idea of pickering emulsions when experimenting with paraffin-water emulsions with solid particles like iron oxide and silicon dioxide. They observed that the micron-sized colloids generated a resistant film at the interface between the two immiscible phases, inhibiting the coalescence of the emulsion drops. These Pickering emulsions, as shown in the figure below, are formed from the self-assembly of colloidal particles in two-part liquid systems, such as oil-water systems. The desorption energy, which is directly related to the stability of emulsions depends on the particle size, particle-particle interaction and, of course, particle-water and particle-oil interactions.

A decrease in total free energy was observed to be a result from the assembly of nanoparticles at an oil/water (O/W) interface. This is shown in the following equation in which a particle with radius r at an interface between oil (O) and water (W) results in a decrease of the initial interfacial energy E_0 to E_1; this difference in energy is ΔE_1.

$$E_0 - E_1 = -\frac{\pi r^2}{\gamma_{O/W}}[\gamma_{O/W} - (\gamma_{P/W} - \gamma_{P/O})]^2$$

At a fixed $\gamma_{P/O}$, $\gamma_{P/W}$, and $\gamma_{O/W}$, the equation shows that the stability of a particle assembly is determined by the radius square. When moving to the interface, particles reduce the unfavorable contact between the immiscible fluids and decrease the interfacial energy. The decrease in total free energy for microscopic particles is much larger than that of thermal energy; this results in an effective confinement of large colloids to the interface. They are irreversibly bound to the interface. Nanoscopic particles are restricted to the interface by an energy reduction comparable to thermal energy. Thus, nanoparticles are easily displaced from the interface. A constant particle exchange then occurs at the interface; the rate of this exchange depends on particle size. The thermally activated escape of small particles occurs more often than larger particles. For the equilibrium state of assembly, the total gain in free energy is smaller for smaller particles. Thus, large nanoparticles assemblies are more stable. This size dependence allows nanoparticles to self-assemble at the interface to attain its equilibrium structure. Micrometer- size colloids, on the other hand, may be confined in a non-equilibrium state.

The interfacial tension and the wettability of a particle surface affect the desorption energy. The contact angle θ between the solid and the oil/water interface determines its wettability. As shown in the figure below, a contact angle θ greater than 90° favors a water-in-oil emersion while a contact angle θ less than 90° favors oil-in-water emulsion. These contact angles affect the stability of the emulsion.

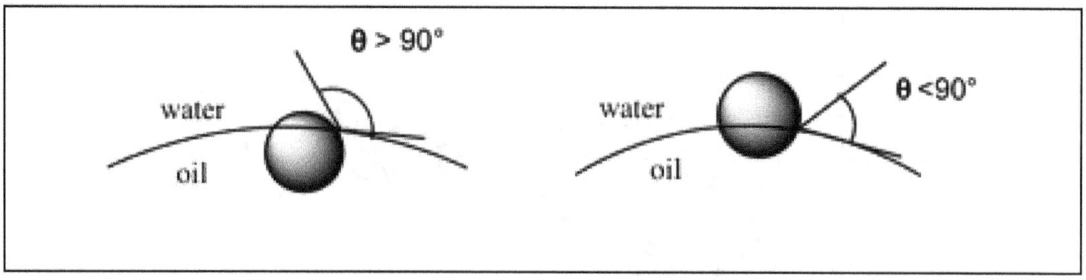

Effects of contact angle on wettability.

A maximum desorption energy peak is observed at a contact angle of 90°. When the contact angle is greater than or less than this point, the desorption energy gradually decreases; thus, the stability of the emulsion decreases as well.

Applications

Material that consists with nano-particles is called "nanostructured material". The phase of "nano-structured material" implies two important ideas:

1. that at least some of the heterogeneity in materials is determined by the size range of nano-structures (about 1–100 nm), and
2. these nanostructures might be synthesized and distributed (or organized), at least in part. The study of self-assembly nanoparticles is important to understand the interaction between single particle in terms of applying them into different applications.

Electronics

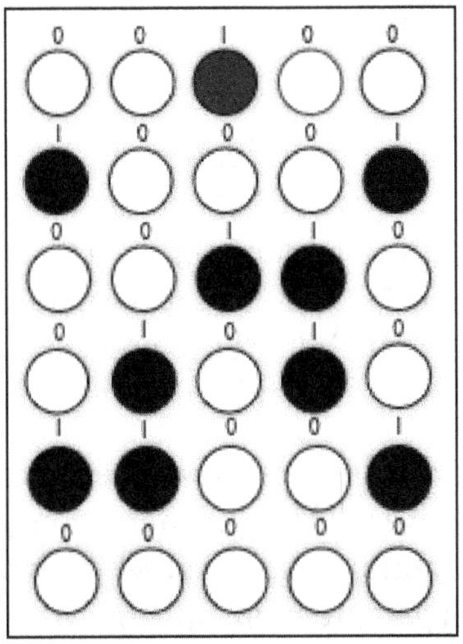

Model of multidimensional array of nano-particles. A particle could have two spins, spin up or down. Based on the spin directions, nano-particles will be able to store 0 and 1. Therefore, nanostructural material has a great potential for future use in electronic devices.

Self-assembly of nanoscale structures from functional nanoparticles has provided a powerful path to developing small and powerful electronic components. In the recent decade, it has been confirmed that organic compounds can be conductors or semiconductors. In addition to insulator, one of the opportunities in nanomaterials science is to use organic synthesis and molecular design to make electronically usefully structures starting with organic molecules. The difficulty of applied nanostructure material is nanoscale objects have always been difficult to work with because they cannot be characterized by molecular techniques and they are too small to see optically. 2D self-assembly monodisperse particle colloids has a strong potential in dense magnetic storage media. Each colloid particle has the ability to store information as known as binary number 0 and 1 after applying it to a strong magnetic field. In the meantime, it requires a nanoscale sensor or detector in order to selectively choose the colloid particle.

Biological Applications

Drug Delivery

Block copolymers offer the ability to self-assemble into uniform, nanosized micelles and accumulate in tumors via the enhanced permeability and retention effect. Polymer composition can be chosen to control the micelle size and compatibility with the drug of choice. The challenges of this application are the difficulty of reproducing or controlling the size of self-assembly nano micelle, preparing predictable size-distribution, and the stability of the micelle with high drug load content.

Magnetic Drug Delivery

Magnetic nanochains are a class of new magnetoresponsive and superparamagnetic nanostructures with highly anisotropic shapes (chain-like) which can be manipulated using magnetic field and magnetic field gradient. The magnetic nanochains possess attractive properties which are significant added value for many potential uses including magneto-mechanical actuation-associated nanomedicines in low and super-low frequency alternating magnetic field and magnetic drug delivery.

Self-assembled RNA triple-helix Nanoparticles Doped on Hydrogels for Cancer Therapy

In 2015, Conde et al. developed a self-assembled dual-color RNA triple-helix structure comprising two miRNAs, a miR mimic (tumor suppressor miRNA) and an antagomiR (oncomiR inhibitor) provides outstanding capability to synergistically abrogate tumors. The authors hypothesized that efficacious delivery of the RNA triple-helix hydrogel nanoconjugates would be achieved by coating the breast tumor with the adhesive hydrogel scaffold that they have previously shown able to enhance the stability of embedded nanoparticles used for local gene and drug delivery using smart gold nanobeacons.

Conjugation of RNA triple-helices to dendrimers allows the formation of stable triplex nanoparticles, which form an RNA-triple-helix adhesive scaffold on interaction with dextran aldehyde, the latter able to chemically interact and adhere to natural tissue amines in the tumor. The authors also show that the self-assembled RNA-triple-helix conjugates remain functional *in vitro* and *in vivo*, and that they lead to nearly 90% levels of tumor shrinkage two weeks post-gel implantation in a triple-negative breast-cancer mouse model. These findings suggest that the RNA-triple-helix hydrogels can be used as an efficient anticancer platform to locally modulate the expression of endogenous miRs in cancer.

The authors reported on a novel strategy for concomitant oncomiR inhibition and tumor suppressor miR replacement therapy using a RNA triple-helix hydrogel scaffold that affords highly efficacious local anticancer therapy. Self-assembled RNA triple-helix conjugates remain functional *in vitro* with high selective uptake and control over miR expression compared to their respective single-stranded or double-stranded forms. Hence, cancer gene delivery systems should provide potent, selective and specific treatment to tumor cells only, unlike the standard delivery of most conventional chemotherapeutic drugs. This approach can be implemented to design self-assembled triplex structures from any other miR combination, or from other genetic materials including antisense DNA or siRNA to treat a range of diseases.

Nanofabrics

Nanofabrics are textiles engineered with small particles that give ordinary materials advantageous properties such as superhydrophobicity, odor and moisture elimination, increased elasticity and strength, and bacterial resistance. Depending on the desired property, a nanofabric is either constructed from nanoscopic fibers called nanofibers, or is formed by applying a solution containing nanoparticles to a regular fabric. Nanofabrics research is an interdisciplinary effort involving bioengineering, molecular chemistry, physics, electrical engineering, computer science, and systems engineering. Applications of nanofabrics have the potential to revolutionize textile manufacturing and areas of medicine such as drug delivery and tissue engineering.

Introduction

A fiber that has a width of less than 1000 nanometers (1000 nm or 1 μm) is generally defined as a nanofiber. A nanoparticle is defined as a small group of atoms or molecules with a radius of less than 100 nanometers (100 nm). Particles on the nanoscale have a very high surface area to volume ratio, whereas this ratio is much lower for objects on the macroscopic scale. A high relative surface area means that a large proportion of a particle's mass exists on its surface, so nanofibers and nanoparticles show a greater level of interaction with other materials. The high surface area to volume ratio observed in very small particles is what makes it possible to create many special properties exhibited by nanofabrics.

The use of nanoparticles and nanofibers to produce specialized nanofabrics became a subject of interest after the sol-gel and electrospinning techniques were fully developed in the 1980s. Since 2000, dramatic increases in global funding have accelerated research efforts in nanotechnology, including nanofabrics research.

Sol-gel

The sol-gel process is used to create gel-like solutions which can be applied to textiles as a liquid finish to create nanofabrics with novel properties. The process begins with dissolving nanoparticles in a liquid solvent (often an alcohol). Once dissolved, several chemical reactions take place that cause the nanoparticles to grow and establish a network throughout the liquid. The network transforms the solution into a colloid (a suspension of solid particles in a liquid) with a gelatinous texture. Finally, the colloid must go through a drying process to remove excess solvent from the mixture before it can be used to treat fabrics. The sol-gel process is used in a similar fashion to make polymer nanofibers, which are long, ultra-thin chains of proteins bonded together.

Electrospinning

Electrospinning extracts nanofibers from polymer solutions (synthesized by the sol-gel process) and collects them to form nonwoven nanofabrics. A strong electric field is applied to the solution to charge the polymer strands. The solution is put into a syringe and aimed at an oppositely charged collector plate. When the force of attraction between the polymer nanofibers and the collector plate exceed the surface tension of the solution, the nanofibers are released from the solution and deposit onto the collector plate. The deposited fibers form a porous nanofabric that can aid in drug delivery and tissue engineering depending on the type of polymer used.

Applications

Textile Manufacturing

When nanoengineered coatings are applied to fabrics, the nanoparticles readily form bonds with the fibers of the material. The high surface area relative to the volume of particles increases their chemical reactivity, allowing them to stick to materials more permanently. Fabrics treated with nanoparticle coatings during manufacturing produce materials that kill bacteria, eliminate moisture and odor, and prevent static electricity. Polymer nanofiber coatings applied to textiles bond to the material at one end of the polymer, forming a surface of tiny, hair-like structures. The polymer "hairs" create a thin layer that prevents liquids from making contact with the actual fabric. Nanofabrics with dirt-proof, stain-proof, and superhydrophobic properties are possible as a result of the layer formed by polymer nanofibers.

Development of nanofabrics for use in the clothing and textiles industry is still in its early stages. Some applications such as bacteria-resistant clothing are not yet practical from an economic standpoint. For example, a Cornell University student's prototype for a bactericidal jacket cost $10,000 alone, so it may be a long time before nanofabric clothing is on the market.

Drug Delivery

Nanofabrics used in medicine can deliver antibiotics, anticancer drugs, proteins, and DNA in precise quantities. Electrospinning creates porous nanofabrics that can be loaded with the desired drug which are then applied to the tissue of the targeted area. The drug passes through the tissue by diffusion, a process in which substances move through a membrane from high to low concentration. The rate at which the drug is administered can be changed by altering the composition of the nanofabric.

Tissue Engineering

Nonwoven fabrics made by electrospinning have the potential to assist in the growth of organ tissue, bone, neurons, tendons, and ligaments. Polymer nanofabrics can act either as a scaffold to support damaged tissue or as a synthetic substitute for actual tissue. Depending on the function, the nanofabric can be made of natural or synthetic polymers, or a combination of both.

Environmental Implications

As nanotechnology advances, many studies have been conducted to determine the effects nano-engineered materials can have on the environment. Most textiles can lose up to 20% of their mass during their lifetime, so nanoparticles used in production of nanofabrics are at risk of being released into the air and waterways.

Nano-silver is expected to have as much as 49.5% of its global production taken by the nanotextiles industry due to its antibacterial properties. It is predicted that 20% of the nano-silver used in the nanofabrics industry will be released into waterways which could cause harm to microorgansims. However, more than 90% of nano-silver is removed during treatment at wastewater facilities, so it is likely that the environmental impact will be minimal. A study on aluminum oxide nanoparticles showed that inhalation caused inflammation in rat lungs. Aluminum oxide nanoparticles are not used in large quantity, so its health risks are negligible. Other studies conducted for nanoparticles suggest that their environmental impact should be low as the nanotextiles industry continues to grow.

Nanoremediation

Nanoremediation is the use of nanoparticles for environmental remediation. It is being explored to treat ground water, wastewater, soil, sediment, or other contaminated environmental materials. Nanoremediation is an emerging industry; by 2009, nanoremediation technologies had been documented in at least 44 cleanup sites around the world, predominantly in the United States. In Europe, nanoremediation is being investigated by the EC funded NanoRem Project. A report produced by the NanoRem consortium has identified around 70 nanoremediation projects worldwide at pilot or full scale. During nanoremediation, a nanoparticle agent must be brought into contact with the target contaminant under conditions that allow a detoxifying or immobilizing reaction. This process typically involves a pump-and-treat process or *in situ* application.

Some nanoremediation methods, particularly the use of nano zero-valent iron for groundwater cleanup, have been deployed at full-scale cleanup sites. Other methods remain in research phases.

Applications

Nanoremediation has been most widely used for groundwater treatment, with additional extensive research in wastewater treatment. Nanoremediation has also been tested for soil and sediment cleanup. Even more preliminary research is exploring the use of nanoparticles to remove toxic materials from gases.

Groundwater Remediation

Currently, groundwater remediation is the most common commercial application of nanoremediation technologies. Using nanomaterials, especially zero-valent metals (ZVMs), for groundwater remediation is an emerging approach that is promising due to the availability and effectiveness of many nanomaterials for degrading or sequestering contaminants.

Nanotechnology offers the potential to effectively treat contaminants *in situ*, avoiding excavation or the need to pump contaminated water out of the ground. The process begins with nanoparticles being injected into a contaminated aquifer via an injection well. The nanoparticles are then transported by groundwater flow to the source of contamination. Upon contact, nanoparticles can sequester contaminants (via adsorption or complexation), immobilizing them, or they can degrade the contaminants to less harmful compounds. Contaminant transformations are typically redox reactions. When the nanoparticle is the oxidant or reductant, it is considered reactive.

The ability to inject nanoparticles to the subsurface and transport them to the contaminant source is imperative for successful treatment. Reactive nanoparticles can be injected into a well where they will then be transported down gradient to the contaminated area. Drilling and packing a well is quite expensive. Direct push wells cost less than drilled wells and are the most often used delivery tool for remediation with nanoiron. A nanoparticle slurry can be injected along the vertical range of the probe to provide treatment to specific aquifer regions.

Surface Water Treatment

The use of various nanomaterials, including carbon nanotubes and TiO_2, shows promise for treatment of surface water, including for purification, disinfection, and desalination. Target contami nants in surface waters include heavy metals, organic contaminants, and pathogens. In this context, nanoparticles may be used as sorbents, as reactive agents (photocatalysts or redox agents), or in membranes used for nanofiltration.

Trace Contaminant Detection

Nanoparticles may assist in detecting trace levels of contaminants in field settings, contributing to effective remediation. Instruments that can operate outside of a laboratory often are not sensitive enough to detect trace contaminants. Rapid, portable, and cost-effective measurement systems for trace contaminants in groundwater and other environmental media would thus enhance contaminant detection and cleanup. One potential method is to separate the analyte from the sample and concentrate them to a smaller volume, easing detection and measurement. When small quantities of solid sorbents are used to absorb the target for concentration, this method is referred to as solid-phase microextraction.

With their high reactivity and large surface area, nanoparticles may be effective sorbents to help concentrate target contaminants for solid-phase microextraction, particularly in the form of self-assembled monolayers on mesoporous supports. The mesoporous silica structure, made through a surfactant templated sol-gel process, gives these self-assembled monolayers high surface area and a rigid open pore structure. This material may be an effective sorbent for many targets, including heavy metals such as mercury, lead, and cadmium, chromate and arsenate, and radionuclides such as ^{99}Tc, ^{137}CS, uranium, and the actinides.

Mechanism

The small size of nanoparticles leads to several characteristics that may enhance remediation. Nanomaterials are highly reactive because of their high surface area per unit mass. Their small particle size also allows nanoparticles to enter small pores in soil or sediment that larger particles might not penetrate, granting them access to contaminants sorbed to soil and increasing the likelihood of contact with the target contaminant.

Because nanomaterials are so tiny, their movement is largely governed by Brownian motion as compared to gravity. Thus, the flow of groundwater can be sufficient to transport the particles. Nanoparticles then can remain suspended in solution longer to establish an *in situ* treatment zone.

Once a nanoparticle contacts the contaminant, it may degrade the contaminant, typically through a redox reaction, or adsorb to the contaminant to immobilize it. In some cases, such as with magnetic nano-iron, adsorbed complexes may be separated from the treated substrate, removing the contaminant. Target contaminants include organic molecules such as pesticides or organic solvents and metals such as arsenic or lead. Some research is also exploring the use of nanoparticles to remove excessive nutrients such as nitrogen and phosphorus.

Materials

A variety of compounds, including some that are used as macro-sized particles for remediation, are being studied for use in nanoremediation. These materials include zero-valent metals like zero-valent iron, calcium carbonate, carbon-based compounds such as graphene or carbon nanotubes, and metal oxides such as titanium dioxide and iron oxide.

Nano Zero-valent Iron

As of 2012, nano zero-valent iron (nZVI) was the nanoscale material most commonly used in bench and field remediation tests. nZVI may be mixed or coated with another metal, such as palladium, silver, or copper, that acts as a catalyst in what is called a bimetallic nanoparticle. nZVI may also be emulsified with a surfactant and an oil, creating a membrane that enhances the nanoparticle's ability to interact with hydrophobic liquids and protects it against reactions with materials dissolved in water. Commercial nZVI particle sizes may sometimes exceed true "nano" dimensions (100 nm or less in diameter).

nZVI appears to be useful for degrading organic contaminants, including chlorinated organic compounds such as polychlorinated biphenyls (PCBs) and trichloroethene (TCE), as well as immobilizing or removing metals. nZVI and other nanoparticles that do not require light can be injected belowground into the contaminated zone for *in situ* groundwater remediation and, potentially, soil remediation.

nZVI nanoparticles can be prepared by using sodium borohydride as the key reductant. $NaBH_4$ (0.2 M) is added into $FeCl_3 \cdot 6H_2$ (0.05 M) solution (~1:1 volume ratio). Ferric iron is reduced via the following reaction:

$$4Fe^{3+} + 3BH_4^- + 9H_2O \rightarrow 4Fe^0 + 3H_2BO_3^- + 12H^+ + 6H_2$$

Palladized Fe particles are prepared by soaking the nanoscale iron particles with an ethanol solution of 1wt% of palladium acetate ($[Pd(C_2H_3O_2)2]_3$). This causes the reduction and deposition of Pd on the Fe surface:

$$Pd^{2+} + Fe^o \rightarrow Pd^o + Fe^{2+}$$

Similar methods may be used to prepared Fe/Pt, Fe/Ag, Fe/Ni, Fe/Co, and Fe/Cu bimetallic particles. With the above methods, nanoparticles of diameter 50-70 nm may be produced. The average specific surface area of Pd/Fe particles is about 35 m^2/g. Ferrous iron salt has also been successfully used as the precursor.

Titanium Dioxide

Titanium dioxide (TiO_2) is also a leading candidate for nanoremediation and wastewater treatment, although as of 2010 it is reported to have not yet been expanded to full-scale commercialization. When exposed to ultraviolet light, such as in sunlight, titanium dioxide produces hydroxyl radicals, which are highly reactive and can oxidize contaminants. Hydroxyl radicals are used for water treatment in methods generally termed advanced oxidation processes. Because light is required for this reaction, TiO_2 is not appropriate for underground *in situ* remediation, but it may be used for wastewater treatment or pump-and-treat groundwater remediation.

TiO_2 is inexpensive, chemically stable, and insoluble in water. TiO_2 has a wide band gap energy (3.2 eV) that requires the use of UV light, as opposed to visible light only, for photocatalytic activation.

To enhance the efficiency of its photocatalysis, research has investigated modifications to TiO_2 or alternative photocatalysts that might use a greater portion of photons in the visible light spectrum. Potential modifications include doping TiO_2 with metals, nitrogen, or carbon.

Challenges

When using *in situ* remediation the reactive products must be considered for two reasons. One reason is that a reactive product might be more harmful or mobile than the parent compound. Another reason is that the products can affect the effectiveness and/or cost of remediation. TCE (trichloroethylene), under reducing conditions by nanoiron, may sequentially dechlorinate to DCE (dichloroethene) and VC (vinyl chloride). VC is known to be more harmful than TCE, meaning this process would be undesirable.

Nanoparticles also react with non-target compounds. Bare nanoparticles tend to clump together and also react rapidly with soil, sediment, or other material in ground water. For *in situ* remediation, this action inhibits the particles from dispersing in the contaminated area, reducing their effectiveness for remediation. Coatings or other treatment may allow nanoparticles to disperse farther and potentially reach a greater portion of the contaminated zone. Coatings for nZVI include surfactants, polyelectrolyte coatings, emulsification layers, and protective shells made from silica or carbon.

Such designs may also affect the nanoparticles' ability to react with contaminants, their uptake by organisms, and their toxicity. A continuing area of research involves the potential for nanoparticles used for remediation to disperse widely and harm wildlife, plants, or people.

In some cases, bioremediation may be used deliberately at the same site or with the same material as nanoremediation. Ongoing research is investigating how nanoparticles may interact with simultaneous biological remediation.

References

- Forrester, D. M.; et al. (2016). "Experimental verification of nanofluid shear-wave reconversion in ultrasonic fields". Nanoscale. doi:10.1039/C5NR07396K
- "Current Intelligence Bulletin 63: Occupational Exposure to Titanium Dioxide". U.S. National Institute for Occupational Safety and Health. April 2011. pp. v–vii, 73–78. Retrieved 27, April 2020
- Mahendran, V. (2012). "Nanofluid based opticalsensor for rapid visual inspection of defects in ferromagnetic materials". Appl. Phys. Lett. 100: 073104. doi:10.1063/1.3684969
- Reeves, Daniel B.; Weaver, John B. (15 December 2012). "Simulations of magnetic nanoparticle Brownian motion". Journal of Applied Physics. 112 (12): 124311. ISSN 0021-8979. doi:10.1063/1.477032
- Buongiorno, J. (March 2006). "Convective Transport in Nanofluids". Journal of Heat Transfer. American Society Of Mechanical Engineers. 128 (3): 240. doi:10.1115/1.2150834. Retrieved 27, March 2020
- Rotello, Vincent (2004). Nanoparticles Building Blocks for Nanotechnology. New York: Springer Science + Business Media Inc. ISBN 0306482878
- Prasad, Paras N. (2012-05-11). Introduction to Nanomedicine and Nanobioengineering. John Wiley & Sons. pp. 121–124. ISBN 9781118351079
- "Radiation Safety Aspects of Nanotechnology". National Council on Radiation Protection and Measurements. 2017-03-02. Retrieved 07, July 2020

- Grzelczak, S.; Vermant, J; Furst, E.M.; Liz-Marzan, L.M. (2010). "Directed Self-Assembly of Nanoparticles". ACS Nano. 4 (7): 3591–3605. doi:10.1021/nn100869j
- Salata, OV (30 April 2004). "Applications of Nanoparticles in Biology and Medicine". Journal of Nanobiotechnology. 2 (1): 3. PMC 419715. PMID 15119954. doi:10.1186/1477-3155-2-3. Retrieved 29, April 2020
- Wang, "Transmission Electron Microscopy of Shape-Controlled Nanocrystals and Their Assemblies", Journal of Physical Chemistry B. 104. (2000): 1153-1175. DOI 10.1021/jp993593c
- Kuznetsov, A.V.; Nield, D.A. "Natural convective boundary-layer flow of a nanofluid past a vertical plate". International Journal of Thermal Sciences. 49 (2): 243–247. doi:10.1016/j.ijthermalsci.2009.07.015
- Das, Sarit K.; Stephen U. S. Choi; Wenhua Yu; T. Pradeep (2007). Nanofluids: Science and Technology. Wiley-Interscience. p. 397. Retrieved 27, March 2020
- Reese, Torsten; Gómez-Vallejo, Vanessa; Ferreira, Paola; Llop, Jordi (2016-03-22). "Health and Safety Considerations for Radiolabelled Nanoparticles". In Llop, Jordi; Gomez-Vallejo, Vanessa; Gibson, Peter Neil. Isotopes in Nanoparticles. Pan Stanford. pp. 493–512. ISBN 9789814669085. doi:10.1201/b19950-19
- Whitesides, G.M.; Kriebel, J.K.; Mayers, B.T. (2005). "Self-Assembly and Nanostructured Materials" (PDF). Nanoscale Assembly – Chemical Techniques. Springer. ISBN 978-0-387-23608-7. doi:10.1007/b106927
- Li, Fei; et al. (2008). "Structural and luminescent properties of ZnO nanorods and ZnO/ZnS nanocomposites". Journal of Alloys and Compounds. 474: 531–535. doi:10.1016/j.jallcom.2008.06.149
- Martín, Isabel García; Frigell, Jens; Llop, Jordi; Marradi, Marco (2016-03-22). "Radiolabelling of NPs Using Radiometals: ^{99m}Tc, ^{68}Ga, ^{67}Ga, ^{89}Zr, and ^{64}Cu". In Llop, Jordi; Gomez-Vallejo, Vanessa; Gibson, Peter Neil. Isotopes in Nanoparticles. Pan Stanford. pp. 183–229. ISBN 9789814669085. doi:10.1201/b19950-9
- Lubick, Naomi (2008-03-01). "Risks of Nanotechnology Remain Uncertain". Environmental Science & Technology. 42 (6): 1821–1824. ISSN 0013-936X. doi:10.1021/es087058e. Retrieved 25, July 2020
- Bardan, R; Lal S; Joshi, A; Halas, Nj (May 2011). "Theranostic Nanoshells: From Probe Design to Imaging and Treatment of Cancer". Accounts of Chemical Research. 44 (10): 936–946. doi:10.1021/ar200023x

4

Common Nanocomposites

The multiphase solids which have one or more dimensions less than 100 nanometers are known as nanocomposites. A few of the major types of nanocomposites are polymer nanocomposite, carbon nanotube, silver nanoparticle and nanowire. All these diverse types of nanocomposites have been carefully analyzed in this chapter.

Polymer Nanocomposite

Polymer nanocomposites (PNC) consist of a polymer or copolymer having nanoparticles or nano-fillers dispersed in the polymer matrix. These may be of different shape (e.g., platelets, fibers, spheroids), but at least one dimension must be in the range of 1–50 nm. These PNC's belong to the category of multi-phase systems (MPS, viz. blends, composites, and foams) that consume nearly 95% of plastics production. These systems require controlled mixing/compounding, stabilization of the achieved dispersion, orientation of the dispersed phase, and the compounding strategies for all MPS, including PNC, are similar.

Polymer nanoscience is the study and application of nanoscience to polymer-nanoparticle matrices, where nanoparticles are those with at least one dimension of less than 100 nm.

The transition from micro- to nano-particles lead to change in its physical as well as chemical properties. Two of the major factors in this are the increase in the ratio of the surface area to volume, and the size of the particle. The increase in surface area-to-volume ratio, which increases as the particles get smaller, leads to an increasing dominance of the behavior of atoms on the surface area of particle over that of those interior of the particle. This affects the properties of the particles when they are reacting with other particles. Because of the higher surface area of the nano-particles, the interaction with the other particles within the mixture is more and this increases the strength, heat resistance, etc. and many factors do change for the mixture.

An example of a nanopolymer is silicon nanospheres which show quite different characteristics; their size is 40–100 nm and they are much harder than silicon, their hardness being between that of sapphire and diamond.

Bio-hybrid Polymer Nanofibers

Many technical applications of biological objects like proteins, viruses or bacteria such as chromatography, optical information technology, sensorics, catalysis and drug delivery require their immobilization. Carbon nanotubes, gold particles and synthetic polymers are used for this purpose. This immobilization has been achieved predominantly by adsorption or by chemical binding and to a lesser extent by incorporating these objects as guests in host matrices. In the guest host systems, an ideal method for the immobilization of biological objects and their integration into hierarchical architectures should be structured on a nanoscale to facilitate the interactions of biological nano-objects with their environment. Due to the large number of natural or synthetic polymers available and the advanced techniques developed to process such systems to nanofibres, rods, tubes etc. make polymers a good platform for the immobilization of biological objects.

Bio-Hybrid Nanofibres by Electrospinning

Polymer fibers are, in general, produced on a technical scale by extrusion, i.e., a polymer melt or a polymer solution is pumped through cylindrical dies and spun/drawn by a take-up device. The resulting fibers have diameters typically on the 10-μm scale or above. To come down in diameter into the range of several hundreds of nanometers or even down to a few nanometers, Electrospinning is today still the leading polymer processing technique available. A strong electric field of the order of 103 V/cm is applied to the polymer solution droplets emerging from a cylindrical die. The electric charges, which are accumulated on the surface of the droplet, cause droplet deformation along the field direction, even though the surface tension counteracts droplet evolution. In supercritical electric fields, the field strength overbears the surface tension and a fluid jet emanates from the droplet tip. The jet is accelerated towards the counter electrode. During this transport phase, the jet is subjected to strong electrically driven circular bending motions that cause a strong elongation and thinning of the jet, a solvent evaporation until, finally, the solid nanofibre is deposited on the counter electrode.

Bio-Hybrid Polymer Nanotubes by Wetting

Electro spinning, co-electrospinning, and the template methods based on nanofibres yield nano-objects which are, in principle, infinitively long. For a broad range of applications including catalysis, tissue engineering, and surface modification of implants this infinite length is an advantage. But in some applications like inhalation therapy or systemic drug delivery, a well-defined length is required. The template method to be described in the following has the advantage such that it allows the preparation of nanotubes and nanorods with very high precision. The method is based on the use of well defined porous templates, such as porous aluminum or silicon.

The basic concept of this method is to exploit wetting processes. A polymer melt or solution is brought into contact with the pores located in materials characterized by high energy surfaces such as aluminum or silicon. Wetting sets in and covers the walls of the pores with a thin film with a thickness of the order of a few tens of nanometers.

Gravity does not play a role, as it is obvious from the fact that wetting takes place independent of the orientation of the pores relative to the direction of gravity. The exact process is still not understood theoretically in detail but its known from experiments that low molar mass systems tend to fill the pores completely, whereas polymers of sufficient chain length just cover the walls. This process happens typically within a minute for temperatures about 50 K above the melting temperature or glass transition temperature, even for highly viscous polymers, such as, for instance, polytetrafluoroethylene, and this holds even for pores with an aspect ratio as large as 10,000. The complete filling, on the other hand, takes days. To obtain nanotubes, the polymer/template system is cooled down to room temperature or the solvent is evaporated, yielding pores covered with solid layers. The resulting tubes can be removed by mechanical forces for tubes up to 10 μm in length, i.e., by just drawing them out from the pores or by selectively dissolving the template. The diameter of the nanotubes, the distribution of the diameter, the homogeneity along the tubes, and the lengths can be controlled.

Applications

The nanofibres, hollow nanofibres, core–shell nanofibres, and nanorods or nanotubes produced have a great potential for a broad range of applications including homogeneous and heterogeneous catalysis, sensorics, filter applications, and optoelectronics. Here we will just consider a limited set of applications related to life science.

Tissue Engineering

This is mainly concerned with the replacement of tissues which have been destroyed by sickness or accidents or other artificial means. The examples are skin, bone, cartilage, blood vessels and may be even organs. This technique involves providing a scaffold on which cells are added and the scaffold should provide favorable conditions for the growth of the same. Nanofibres have been found to provide very good conditions for the growth of such cells, one of the reasons being that fibrillar structures can be found on many tissues which allow the cells to attach strongly to the fibers and grow along them as shown.

Nanoparticles such as graphene, carbon nanotubes, molybdenum disulfide and tungsten disulfide are being used as reinforcing agents to fabricate mechanically strong biodegradable polymeric nanocomposites for bone tissue engineering applications. The addition of these nanoparticles in the polymer matrix at low concentrations (~0.2 weight %) leads to significant improvements in the compressive and flexural mechanical properties of polymeric nanocomposites. Potentially, these nanocomposites may be used as a novel, mechanically strong, light weight composite as bone implants. The results suggest that mechanical reinforcement is dependent on the nanostructure morphology, defects, dispersion of nanomaterials in the polymer matrix, and the cross-linking density of the polymer. In general, two-dimensional nanostructures can reinforce the polymer better than one-dimensional nanostructures, and inorganic nanomaterials are better reinforcing agents than carbon based nanomaterials.

Delivery from Compartmented Nanotubes

Nano tubes are also used for carrying drugs in general therapy and in tumor therapy in particular. The role of them is to protect the drugs from destruction in blood stream, to control the delivery with a well-defined release kinetics, and in ideal cases, to provide vector-targeting properties or release mechanism by external or internal stimuli.

Rod or tube-like, rather than nearly spherical, nanocarriers may offer additional advantages in terms of drug delivery systems. Such drug carrier particles possess additional choice of the axial ratio, the curvature, and the "all-sweeping" hydrodynamic-related rotation, and they can be modified chemically at the inner surface, the outer surface, and at the end planes in a very selective way. Nanotubes prepared with a responsive polymer attached to the tube opening allow the control of access to and release from the tube. Furthermore, nanotubes can also be prepared showing a gradient in its chemical composition along the length of the tube.

Compartmented drug release systems were prepared based on nanotubes or nanofibres. Nanotubes and nanofibres, for instance, which contained fluorescent albumin with dog-fluorescein isothiocyanate were prepared as a model drug, as well as super paramagnetic nanoparticles composed of iron oxide or nickel ferrite. The presence of the magnetic nanoparticles allowed, first of all, the guiding of the nanotubes to specific locations in the body by external magnetic fields. Super paramagnetic particles are known to display strong interactions with external magnetic fields leading to large saturation magnetizations. In addition, by using periodically varying magnetic fields, the nanoparticles were heated up to provide, thus, a trigger for drug release. The presence of the model drug was established by fluorescence spectroscopy and the same holds for the analysis of the model drug released from the nanotubes.

Immobilization of Proteins

Core shell fibers of nano particles with fluid cores and solid shells can be used to entrap biological objects such as proteins, viruses or bacteria in conditions which do not affect their functions. This effect can be used among others for biosensor applications. For example, Green Fluorescent Protein is immobilized in nanostructured fibres providing large surface areas and short distances for the analyte to approach the sensor protein.

With respect to using such fibers for sensor applications fluorescence of the core shell fibers was found to decay rapidly as the fibers were immersed into a solution containing urea: urea permeates through the wall into the core where it causes denaturation of the GFP. This simple experiment reveals that core–shell fibers are promising objects for preparing biosensors based on biological objects.

Polymer nanostructured fibers, core–shell fibers, hollow fibers, and nanorods and nanotubes provide a platform for a broad range of applications both in material science as well as in life science. Biological objects of different complexity and synthetic objects carrying specific functions can be incorporated into such nanostructured polymer systems while keeping their specific functions vital. Biosensors, tissue engineering, drug delivery, or enzymatic catalysis is just a few of the possible examples. The incorporation of viruses and bacteria all the way up to microorganism should not really pose a problem and the applications coming from such biohybrid systems should be tremendous.

Size and Pressure Effects on Nanopolymers

The size- and pressure- dependent glass transition temperatures of free-standing films or supported films having weak interactions with **substrates** decreases with decreasing of pressure and size. However, the glass transition temperature of supported films having strong interaction with substrates increases of pressure and the decrease of size. Different models like two layer model, three layer model, T_g (D, 0) 1/D and some more models relating specific heat, density and thermal expansion are used to obtain the experimental results on nanopolymers and even some observations like freezing of films due to memory effects in the visco-elastic eigenmodels of the films, and finite effects of the small molecule glass are observed. To describe T_g (D, 0) function of polymers more generally, a simple and unified model recently is provided based on the size-dependent melting temperature of crystals and Lindemann's criterion

$$T_g(D,0) / T_g(\infty,0)\, \sigma_g^2(\infty,0) / \sigma_g^2(D,0)$$

where σ_g is the root of mean squared displacement of surface and interior molecules of glasses at T_g (D, 0), $\alpha = \sigma_s^2 (D, 0) / \sigma_v^2 (D, 0)$ with subscripts *s* and *υ* denoting surface and volume, respectively. For a nanoparticle, D has a usual meaning of diameter, for a nanowire, D is taken as its diameter, and for a thin film, D denotes its thickness. D_0 denotes a critical diameter at which all molecules of a low-dimensional glass are located on its surface.

Conclusion

The devices that use the properties of low-dimensional objects such as nanoparticles are promising due to the possibility of tailoring a number of electrophysical, optical and magnetic properties changing the size of nanoparticles, which can be controlled during the synthesis. In the case of polymer nanocomposites we can use the properties of disordered systems.

Here recent developments in the field of polymer nano-composites and some of their applications have been reviewed. Though there is much use in this field, there are many limitations also. For example, in the release of drugs using nanofibres, cannot be controlled independently and a burst release is usually the case, whereas a more linear release is required. Let us now consider future aspects in this field.

There is a possibility of building ordered arrays of nanoparticles in the polymer matrix. A number of possibilities also exist to manufacture the nanocomposite circuit boards. An even more attractive method exists to use polymer nanocomposites for neural networks applications. Another promising area of development is optoelectronics and optical computing. The single domain nature and super paramagnetic behavior of nanoparticles containing ferromagnetic metals could be possibly used for magneto-optical storage media manufacturing.

Carbon Nanotube

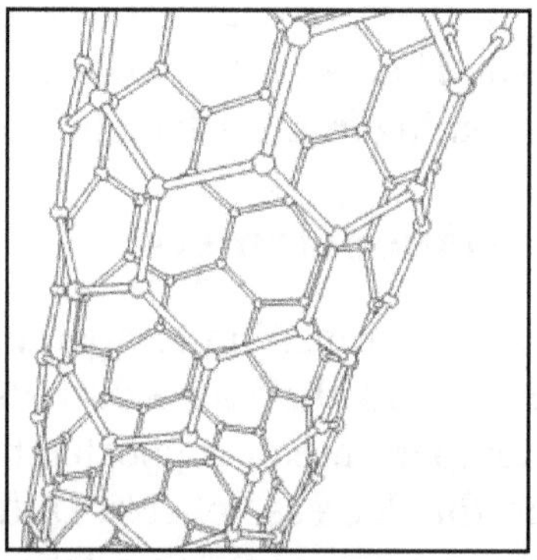

Rotating single-walled zigzag carbon nanotube.

Carbon nanotubes (CNTs) are allotropes of carbon with a cylindrical nanostructure. Nanotubes have been constructed with length-to-diameter ratio of up to 132,000,000:1, significantly larger than for any other material. These cylindrical carbon molecules have unusual properties, which are valuable for nanotechnology, electronics, optics and other fields of materials science and technology. In particular, owing to their extraordinary thermal conductivity and mechanical and electrical properties, carbon nanotubes find applications as additives to various structural materials.

For instance, nanotubes form a tiny portion of the material(s) in some (primarily carbon fiber) baseball bats, golf clubs, car parts or damascus steel.

Nanotubes are members of the fullerene structural family. Their name is derived from their long, hollow structure with the walls formed by one-atom-thick sheets of carbon, called graphene. These sheets are rolled at specific and discrete ("chiral") angles, and the combination of the rolling angle and radius decides the nanotube properties; for example, whether the individual nanotube shell is a metal or semiconductor. Nanotubes are categorized as single-walled nanotubes (SWNTs) and multi-walled nanotubes (MWNTs). Individual nanotubes naturally align themselves into "ropes" held together by van der Waals forces, more specifically, pi-stacking.

Applied quantum chemistry, specifically, orbital hybridization best describes chemical bonding in nanotubes. The chemical bonding of nanotubes is composed entirely of sp^2 bonds, similar to those of graphite. These bonds, which are stronger than the sp^3 bonds found in alkanes and diamond, provide nanotubes with their unique strength.

Types of Carbon Nanotubes and Related Structures

Terminology

There is no consensus on some terms describing carbon nanotubes in scientific literature: both "-wall" and "-walled" are being used in combination with "single", "double", "triple" or "multi", and the letter C is often omitted in the abbreviation; for example, multi-walled carbon nanotube (MWNT).

Single-walled

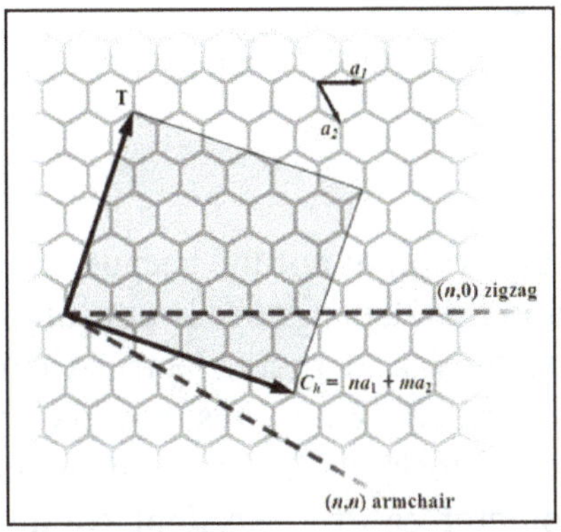

The (n,m) nanotube naming scheme can be thought of as a vector ($\mathbf{C}_h$) in an infinite graphene sheet that describes how.

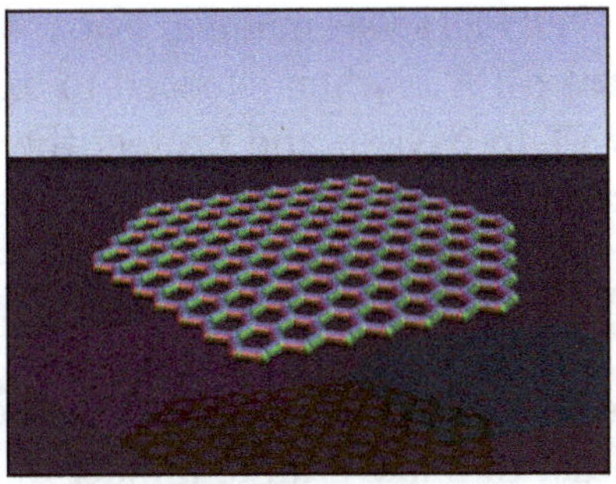

Graphene nanoribbon to "roll up" the graphene sheet to make the nanotube. **T** denotes the tube axis, and $\mathbf{a}_1$ and $\mathbf{a}_2$ are the unit vectors of graphene in real space.

A scanning tunneling microscopy image of single-walled carbon nanotube.

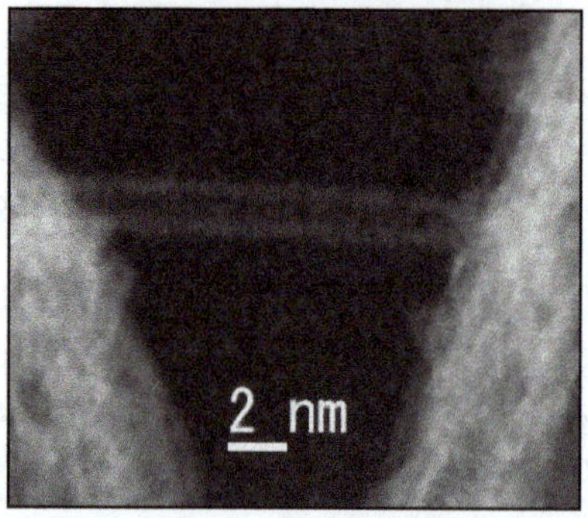

A transmission electron microscopy image of a single-walled carbon nanotube.

Most single-walled nanotubes (SWNTs) have a diameter of close to 1 nanometer, and can be many millions of times longer. The structure of a SWNT can be conceptualized by wrapping a one-atom-thick layer of graphite called graphene into a seamless cylinder. The way the graphene sheet is wrapped is represented by a pair of indices (n,m). The integers n and m denote the number of unit vectors along two directions in the honeycomb crystal lattice of graphene. If $m = 0$, the nanotubes are called zigzag nanotubes, and if $n = m$, the nanotubes are called armchair nanotubes. Otherwise, they are called chiral. The diameter of an ideal nanotube can be calculated from its (n,m) indices as follows

$$d = \frac{a}{\pi}\sqrt{(n^2 + nm + m^2)} = 78.3\sqrt{((n+m)^2 - nm)}\text{pm},$$

where $a = 0.246$ nm.

SWNTs are an important variety of carbon nanotube because most of their properties change significantly with the (n,m) values, and this dependence is non-monotonic. In par-ticular, their band gap can vary from zero to about 2 eV and their electrical conductivity can show metallic or semiconducting behavior. Single-walled nanotubes are likely candidates for miniatur-izing electronics. The most basic building block of these systems is the electric wire, and SWNTs with diameters of an order of a nanometer can be excellent conductors. One useful application of SWNTs is in the development of the first intermolecular field-effect transistors (FET). The first intermolecular logic gate using SWCNT FETs was made in 2001. A logic gate requires both a p-FET and an n-FET. Because SWNTs are p-FETs when exposed to oxygen and n-FETs otherwise, it is possible to expose half of an SWNT to oxygen and protect the other half from it. The resulting SWNT acts as a *not* logic gate with both p and n-type FETs in the same molecule.

Prices for single-walled nanotubes declined from around \$1500 per gram as of 2000 to retail prices of around \$50 per gram of as-produced 40–60% by weight SWNTs as of March 2010. As of 2016 the retail price of as-produced 75% by weight SWNTs were \$2 per gram, cheap enough for widespread use. SWNTs are forecast to make a large impact in electronics applications by 2020 according to the *The Global Market for Carbon Nanotubes* report.

Multi-walled

Multi-walled nanotubes (MWNTs) consist of multiple rolled layers (concentric tubes) of graphene. There are two models that can be used to describe the structures of multi-walled nanotubes. In the *Russian Doll* model, sheets of graphite are arranged in concentric cylinders, e.g., a (0,8) single-walled nanotube (SWNT) within a larger (0,17) single-walled nanotube. In the *Parchment* model, a single sheet of graphite is rolled in around itself, resembling a scroll of parchment or a rolled newspaper. The interlayer distance in multi-walled nanotubes is close to the distance between graphene layers in graphite, approximately 3.4 Å. The Russian Doll structure is observed more commonly. Its individual shells can be described as SWNTs, which can be metallic or semiconducting. Because of statistical probability and restrictions on the relative diameters of the individual tubes, one of the shells, and thus the whole MWNT, is usually a zero-gap metal.

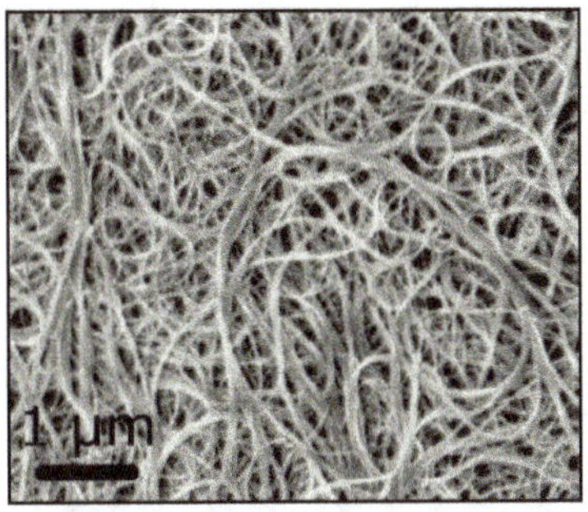

A scanning electron microscopy image of carbon nanotubes bundles.

Double-walled carbon nanotubes (DWNTs) form a special class of nanotubes because their morphology and properties are similar to those of SWNTs but they are more resistant to chemicals. This is especially important when it is necessary to graft chemical functions to the surface of the nanotubes (functionalization) to add properties to the CNT. Covalent functionalization of SWNTs will break some C=C double bonds, leaving "holes" in the structure on the nanotube, and thus modifying both its mechanical and electrical properties. In the case of DWNTs, only the outer wall is modified. DWNT synthesis on the gram-scale was first proposed in 2003 by the CCVD technique, from the selective reduction of oxide solutions in methane and hydrogen.

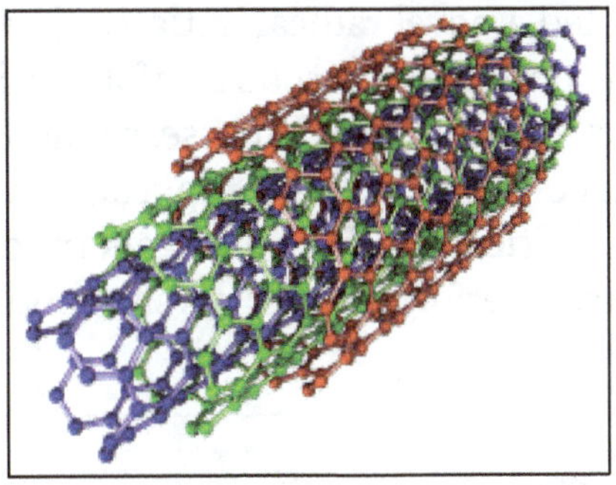

Triple-walled armchair carbon nanotube.

The telescopic motion ability of inner shells and their unique mechanical properties will permit the use of multi-walled nanotubes as main movable arms in coming nanomechanical devices. Retraction force that occurs to telescopic motion caused by the Lennard-Jones interaction between shells and its value is about 1.5 nN.

Torus

In theory, a nanotorus is a carbon nanotube bent into a torus (doughnut shape). Nanotori are predicted to have many unique properties, such as magnetic moments 1000 times larger than previously expected for certain specific radii. Properties such as magnetic moment, thermal stability, etc. vary widely depending on radius of the torus and radius of the tube.

Nanobud

Carbon nanobuds are a newly created material combining two previously discovered allotropes of carbon: carbon nanotubes and fullerenes. In this new material, fullerene-like "buds" are covalently bonded to the outer sidewalls of the underlying carbon nanotube. This hybrid material has useful properties of both fullerenes and carbon nanotubes. In particular, they have been found to be exceptionally good field emitters. In composite materials, the attached fullerene molecules may function as molecular anchors preventing slipping of the nanotubes, thus improving the composite's mechanical properties.

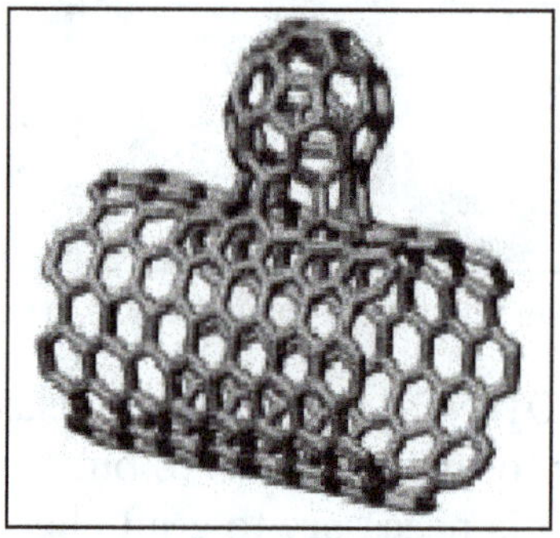

A stable nanobud structure.

Three-dimensional Carbon Nanotube Architectures

Recently, several studies have highlighted the prospect of using carbon nanotubes as building blocks to fabricate three-dimensional macroscopic (>100 nm in all three dimensions) all-carbon devices. Lalwani et al. have reported a novel radical initiated thermal crosslinking method to fabricate macroscopic, free-standing, porous, all-carbon scaffolds using single- and multi-walled carbon nanotubes as building blocks. These scaffolds possess macro-, micro-, and nano- structured pores and the porosity can be tailored for specific applications. These 3D all-carbon scaffolds/architectures may be used for the fabrication of the next generation of energy storage, supercapacitors, field emission transistors, high-performance catalysis, photovoltaics, and biomedical devices and implants.

3D carbon scaffolds.

Graphenated Carbon Nanotubes (g-CNTs)

SEM series of graphenated CNTs with varying foliate density

Graphenated CNTs are a relatively new hybrid that combines graphitic foliates grown along the sidewalls of multiwalled or bamboo style CNTs. Yu *et al.* reported on "chemically bonded graphene leaves" growing along the sidewalls of CNTs. Stoner *et al.* described these structures as "graphenated CNTs" and reported in their use for enhanced supercapacitor performance. Hsu *et al.* further reported on similar structures formed on carbon fiber paper, also for use in supercapacitor applications. Pham *et al.* also reported a similar structure, namely "graphene-carbon nanotube hybrids", grown directly onto carbon fiber paper to form an integrated, binder free, high surface area conductive catalyst support for Proton Exchange Membrane Fuel Cells electrode applications with enhanced performance and durability. The foliate density can vary as a function of deposition conditions (e.g. temperature and time) with their structure ranging from few layers of graphene (< 10) to thicker, more graphite-like.

The fundamental advantage of an integrated graphene-CNT structure is the high surface area three-dimensional framework of the CNTs coupled with the high edge density of graphene. Graphene edges provide significantly higher charge density and reactivity than the basal plane, but they are difficult to arrange in a three-dimensional, high volume-density geometry. CNTs are readily aligned in a high density geometry (i.e., a vertically aligned forest) but lack high charge density surfaces—the sidewalls of the CNTs are similar to the basal plane of graphene and exhibit low charge density except where edge defects exist. Depositing a high density of graphene foliates along the length of aligned CNTs can significantly increase the total charge capacity per unit of nominal area as compared to other carbon nanostructures.

Nitrogen-doped Carbon Nanotubes

Nitrogen doped carbon nanotubes (N-CNTs) can be produced through five main methods, chemical vapor deposition, high-temperature and high-pressure reactions, gas-solid reaction of amorphous carbon with NH_3 at high temperature, solid reaction, and solvothermal synthesis.

N-CNTs can also be prepared by a CVD method of pyrolyzing melamine under Ar at elevated temperatures of 800–980 °C. However synthesis by CVD of melamine results in the formation of bamboo-structured CNTs. XPS spectra of grown N-CNTs reveal nitrogen in five main components, pyridinic nitrogen, pyrrolic nitrogen, quaternary nitrogen, and nitrogen oxides. Furthermore, synthesis temperature affects the type of nitrogen configuration.

Nitrogen doping plays a pivotal role in lithium storage, as it creates defects in the CNT walls allowing for Li ions to diffuse into interwall space. It also increases capacity by providing more favorable bind of N-doped sites. N-CNTs are also much more reactive to metal oxide nanoparticle deposition which can further enhance storage capacity, especially in anode materials for Li-ion batteries. However boron-doped nanotubes have been shown to make batteries with triple capacity.

Peapod

A carbon peapod is a novel hybrid carbon material which traps fullerene inside a carbon nanotube. It can possess interesting magnetic properties with heating and irradiation. It can also be applied as an oscillator during theoretical investigations and predictions.

Cup-stacked Carbon Nanotubes

Cup-stacked carbon nanotubes (CSCNTs) differ from other quasi-1D carbon structures, which normally behave as quasi-metallic conductors of electrons. CSCNTs exhibit semiconducting behaviors due to the stacking microstructure of graphene layers.

Extreme Carbon Nanotubes

The observation of the *longest* carbon nanotubes grown so far are over 1/2 m (550 mm long) was reported in 2013. These nanotubes were grown on Si substrates using an improved chemical vapor deposition (CVD) method and represent electrically uniform arrays of single-walled carbon nanotubes.

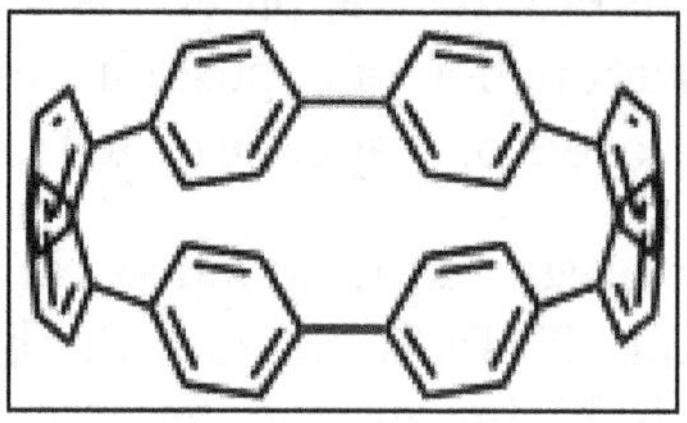

Cycloparaphenylene

The *shortest* carbon nanotube is the organic compound cycloparaphenylene, which was synthesized in early 2009.

The *thinnest* carbon nanotube is the armchair (2,2) CNT with a diameter of 0.3 nm. This nanotube was grown inside a multi-walled carbon nanotube. Assigning of carbon nanotube type was done by a combination of high-resolution transmission electron microscopy (HRTEM), Raman spectroscopy and density functional theory (DFT) calculations.

The *thinnest freestanding* single-walled carbon nanotube is about 0.43 nm in diameter. Researchers suggested that it can be either (5,1) or (4,2) SWCNT, but the exact type of carbon nanotube remains questionable. (3,3), (4,3) and (5,1) carbon nanotubes (all about 0.4 nm in diameter) were unambiguously identified using aberration-corrected high-resolution transmission electron microscopy inside double-walled CNTs.

The *highest density* of CNTs was achieved in 2013, grown on a conductive titanium-coated copper surface that was coated with co-catalysts cobalt and molybdenum at lower than typical temperatures of 450 °C. The tubes averaged a height of 380 nm and a mass density of 1.6 g cm^{-3}. The material showed ohmic conductivity (lowest resistance 22 kΩ).

Properties

Strength

Carbon nanotubes are the strongest and stiffest materials yet discovered in terms of tensile strength and elastic modulus respectively. This strength results from the covalent sp^2 bonds formed between the individual carbon atoms. In 2000, a multi-walled carbon nanotube was tested to have a tensile strength of 63 gigapascals (9,100,000 psi). (For illustration, this translates into the ability to endure tension of a weight equivalent to 6,422 kilograms-force (62,980 N; 14,160 lbf) on a cable with cross-section of 1 square millimetre (0.0016 sq in).)

Further studies, such as one conducted in 2008, revealed that individual CNT shells have strengths of up to ~100 gigapascals (15,000,000 psi), which is in agreement with quantum/atomistic models. Since carbon nanotubes have a low density for a solid of 1.3 to 1.4 g/cm^3, its specific strength of up to 48,000 $kN{\cdot}m{\cdot}kg^{-1}$ is the best of known materials, compared to high-carbon steel's 154 $kN{\cdot}m{\cdot}kg^{-1}$.

Under excessive tensile strain, the tubes will undergo plastic deformation, which means the deformation is permanent. This deformation begins at strains of approximately 5% and can increase the maximum strain the tubes undergo before fracture by releasing strain energy.

Although the strength of individual CNT shells is extremely high, weak shear interactions between adjacent shells and tubes lead to significant reduction in the effective strength of multi-walled carbon nanotubes and carbon nanotube bundles down to only a few GPa. This limitation has been recently addressed by applying high-energy electron irradiation, which crosslinks inner shells and tubes, and effectively increases the strength of these materials to ~60 GPa for multi-walled carbon nanotubes and ~17 GPa for double-walled carbon nanotube bundles.

CNTs are not nearly as strong under compression. Because of their hollow structure and high aspect ratio, they tend to undergo buckling when placed under compressive, torsional, or bending stress.

Comparison of mechanical properties			
Material	**Young's modulus (TPa)**	**Tensile strength (GPa)**	**Elongation at break (%)**
SWNT[E]	~1 (from 1 to 5)	13–53	16
Armchair SWNT[T]	0.94	126.2	23.1
Zigzag SWNT[T]	0.94	94.5	15.6–17.5
Chiral SWNT	0.92		
MWNT[E]	0.2–0.8–0.95	11–63–150	
Stainless steel[E]	0.186–0.214	0.38–1.55	15–50
Kevlar–29&149[E]	0.06–0.18	3.6–3.8	~2

[E]Experimental observation; [T]Theoretical prediction

The above discussion referred to axial properties of the nanotube, whereas simple geometrical considerations suggest that carbon nanotubes should be much softer in the radial direction than along the tube axis. Indeed, TEM observation of radial elasticity suggested that even the van der Waals forces can deform two adjacent nanotubes. Nanoindentation experiments, performed by several groups on multiwalled carbon nanotubes and tapping/contact mode atomic force microscope measurements performed on single-walled carbon nanotubes, indicated a Young's modulus of the order of several GPa, confirming that CNTs are indeed rather soft in the radial direction.

Hardness

Standard single-walled carbon nanotubes can withstand a pressure up to 25 GPa without [plastic/permanent] deformation. They then undergo a transformation to superhard phase nanotubes. Maximum pressures measured using current experimental techniques are around 55 GPa. However, these new superhard phase nanotubes collapse at an even higher, albeit unknown, pressure.

The bulk modulus of superhard phase nanotubes is 462 to 546 GPa, even higher than that of diamond (420 GPa for single diamond crystal).

Wettability

The surface wettability of CNT is of importance for its applications in various settings. Although the intrinsic contact angle of graphite is around 90°, the contact angles of most as-synthesized CNT arrays are over 160°, exhibiting a superhydrophobic property. By applying a low voltage as low as 1.3V, the extreme water repellant surface can be switched into superhydrophilic.

Kinetic Properties

Multi-walled nanotubes are multiple concentric nanotubes precisely nested within one another. These exhibit a striking telescoping property whereby an inner nanotube core may slide, almost without friction, within its outer nanotube shell, thus creating an atomically perfect linear or rotational bearing. This is one of the first true examples of molecular nanotechnology, the precise positioning of atoms to create useful machines. Already, this property has been utilized to create the world's smallest rotational motor. Future applications such as a gigahertz mechanical oscillator are also envisioned.

Electrical Properties

Unlike graphene, which is a two-dimensional semimetal, carbon nanotubes are either metallic or semiconducting along the tubular axis. For a given (n,m) nanotube, if $n = m$, the nanotube is metallic; if $n - m$ is a multiple of 3, then the nanotube is semiconducting with a very small band gap, otherwise the nanotube is a moderate semiconductor. Thus all armchair $(n = m)$ nanotubes are metallic, and nanotubes (6,4), (9,1), etc. are semiconducting. Carbon nanotubes are not semimetallic because the degenerate point (that point where the π [bonding] band meets the π^* [anti-bonding] band, at which the energy goes to zero) is slightly shifted away from the K point in the Brillouin zone due to the curvature of the tube surface, casing hybridization between the σ^* and π^* anti-bonding bands, modifying the band dispersion.

The rule regarding metallic versus semiconductor behavior has exceptions, because curvature effects in small diameter tubes can strongly influence electrical properties. Thus, a (5,0) SWCNT that should be semiconducting in fact is metallic according to the calculations. Likewise, zigzag and chiral SWCNTs with small diameters that should be metallic have a finite gap (armchair nanotubes remain metallic). In theory, metallic nanotubes can carry an electric current density of 4×10^9 A/cm^2, which is more than 1,000 times greater than those of metals such as copper, where for copper interconnects current densities are limited by electromigration. Carbon nanotubes are thus being explored as conductivity enhancing components in composite materials and many groups are attempting to commercialize highly conducting electrical wire assembled from individual carbon nanotubes. There are significant challenges to be overcome, however, such as the much more resistive nanotube-to-nanotube junctions and impurities, all of which lower the electrical conductivity of the macroscopic nanotube wires by orders of magnitude, as compared to the conductivity of the individual nanotubes.

Because of its nanoscale cross-section, electrons propagate only along the tube's axis. As a result, carbon nanotubes are frequently referred to as one-dimensional conductors. The maximum electrical conductance of a single-walled carbon nanotube is $2G_0$, where $G_0 = 2e^2/h$ is the conductance of a single ballistic quantum channel.

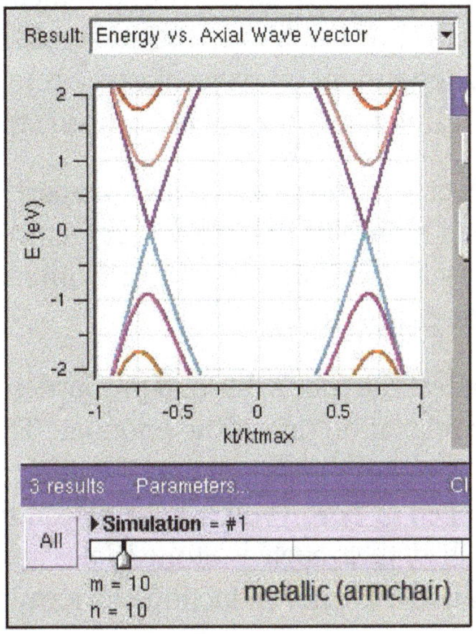

Band structures computed using tight binding approximation for (6,0) CNT (zigzag, metallic), (10,2) CNT (semiconducting) and (10,10) CNT (armchair, metallic).

Due to the role of the π-electron system in determining the electronic properties of graphene, doping in carbon nanotubes differs from that of bulk crystalline semiconductors from the same group of the periodic table (e.g. silicon). Graphitic substitution of carbon atoms in the nanotube wall by boron or nitrogen dopants leads to p-type and n-type behavior, respectively, as would be expected in silicon. However, some non-substitutional (intercalated or adsorbed) dopants introduced into a carbon nanotube, such as alkali metals as well as electron-rich metallocenes, result in n-type conduction because they donate electrons to the π-electron system of the nanotube. By contrast, π-electron acceptors such as $FeCl_3$ or electron-deficient metallocenes function as p-type dopants since they draw π-electrons away from the top of the valence band.

Intrinsic superconductivity has been reported, although other experiments found no evidence of this, leaving the claim a subject of debate.

Optical Properties

Thermal Properties

All nanotubes are expected to be very good thermal conductors along the tube, exhibiting a property known as "ballistic conduction", but good insulators lateral to the tube axis. Measurements show that a SWNT has a room-temperature thermal conductivity along its axis of about 3500 $W·m^{-1}·K^{-1}$; compare this to copper, a metal well known for its good thermal conductivity, which transmits 385 $W·m^{-1}·K^{-1}$. A SWNT has a room-temperature thermal conductivity across its axis (in the radial direction) of about 1.52 $W·m^{-1}·K^{-1}$, which is about as thermally conductive as soil. The temperature stability of carbon nanotubes is estimated to be up to 2800 °C in vacuum and about 750 °C in air.

Defects

As with any material, the existence of a crystallographic defect affects the material properties. Defects can occur in the form of atomic vacancies. High levels of such defects can lower the tensile strength by up to 85%. An important example is the Stone Wales defect, which creates a pentagon and heptagon pair by rearrangement of the bonds. Because of the very small structure of CNTs, the tensile strength of the tube is dependent on its weakest segment in a similar manner to a chain, where the strength of the weakest link becomes the maximum strength of the chain.

Crystallographic defects also affect the tube's electrical properties. A common result is lowered conductivity through the defective region of the tube. A defect in armchair-type tubes (which can conduct electricity) can cause the surrounding region to become semiconducting, and single monatomic vacancies induce magnetic properties.

Crystallographic defects strongly affect the tube's thermal properties. Such defects lead to phonon scattering, which in turn increases the relaxation rate of the phonons. This reduces the mean free path and reduces the thermal conductivity of nanotube structures. Phonon transport simulations indicate that substitutional defects such as nitrogen or boron will primarily lead to scattering of high-frequency optical phonons. However, larger-scale defects such as Stone Wales defects cause phonon scattering over a wide range of frequencies, leading to a greater reduction in thermal conductivity.

Safety and Health

The toxicity of carbon nanotubes has been an important question in nanotechnology. As of 2007, such research had just begun. The data is still fragmentary and subject to criticism. Preliminary results highlight the difficulties in evaluating the toxicity of this heterogeneous material. Parameters such as structure, size distribution, surface area, surface chemistry, surface charge, and agglomeration state as well as purity of the samples, have considerable impact on the reactivity of carbon nanotubes. However, available data clearly show that, under some conditions, nanotubes can cross membrane barriers, which suggests that, if raw materials reach the organs, they can induce harmful effects such as inflammatory and fibrotic reactions.

Synthesis

Techniques have been developed to produce nanotubes in sizable quantities, including arc discharge, laser ablation, high-pressure carbon monoxide disproportionation, and chemical vapor deposition (CVD). Most of these processes take place in a vacuum or with process gases. CVD growth of CNTs can occur in vacuum or at atmospheric pressure. Large quantities of nanotubes can be synthesized by these methods; advances in catalysis and continuous growth are making CNTs more commercially viable.

Chemical Modification

Carbon nanotubes can be functionalized to attain desired properties that can be used in a wide variety of applications. The two main methods of carbon nanotube functionalization are covalent and non-covalent modifications. Because of their hydrophobic nature, carbon nanotubes tend to agglomerate hindering their dispersion is solvents or viscous polymer melts. The resulting nanotube bundles or aggregates reduce the mechanical performance of the final composite. The surface of the carbon nanotubes can be modified to reduce the hydrophobicity and improve interfacial adhesion to a bulk polymer through chemical attachment.

Applications

Current

Current use and application of nanotubes has mostly been limited to the use of bulk nanotubes, which is a mass of rather unorganized fragments of nanotubes. Bulk nanotube materials may never achieve a tensile strength similar to that of individual tubes, but such composites may, nevertheless, yield strengths sufficient for many applications. Bulk carbon nanotubes have already been used as composite fibers in polymers to improve the mechanical, thermal and electrical properties of the bulk product.

- Easton-Bell Sports, Inc. have been in partnership with Zyvex Performance Materials, using CNT technology in a number of their bicycle components—including flat and riser handlebars, cranks, forks, seatposts, stems and aero bars.
- Zyvex Technologies has also built a 54' maritime vessel, the Piranha Unmanned Surface Vessel, as a technology demonstrator for what is possible using CNT technology. CNTs help improve the structural performance of the vessel, resulting in a lightweight 8,000 lb boat that can carry a payload of 15,000 lb over a range of 2,500 miles.
- Amroy Europe Oy manufactures Hybtonite carbon nanoepoxy resins where carbon nanotubes have been chemically activated to bond to epoxy, resulting in a composite material that is 20% to 30% stronger than other composite materials. It has been used for wind turbines, marine paints and variety of sports gear such as skis, ice hockey sticks, baseball bats, hunting arrows, and surfboards.

Other current applications include:

- tips for atomic force microscope probes
- in tissue engineering, carbon nanotubes can act as scaffolding for bone growth

There is also ongoing research in using carbon nanotubes as a scaffold for diverse microfabrication techniques.

Potential

The strength and flexibility of carbon nanotubes makes them of potential use in controlling other nanoscale structures, which suggests they will have an important role in nanotechnology engineering. The highest tensile strength of an individual multi-walled carbon nanotube has been tested to be 63 GPa. Carbon nanotubes were found in Damascus steel from the 17th century, possibly helping to account for the legendary strength of the swords made of it. Recently, several studies have highlighted the prospect of using carbon nanotubes as building blocks to fabricate three-dimensional macroscopic (>1mm in all three dimensions) all-carbon devices. Lalwani et al. have reported a novel radical initiated thermal crosslinking method to fabricated macroscopic, free-standing, porous, all-carbon scaffolds using single- and multi-walled carbon nanotubes as building blocks. These scaffolds possess macro-, micro-, and nano- structured pores and the porosity can be tailored for specific applications. These 3D all-carbon scaffolds/architectures maybe used for the fabrication of the next generation of energy storage, supercapacitors, field emission transistors, high-performance catalysis, photovoltaics, and biomedical devices and implants.

Discovery

The true identity of the discoverers of carbon nanotubes is a subject of some controversy. For years, scientists assumed that Sumio Iijima of NEC had discovered carbon nanotubes in 1991. He published a paper describing his discovery which initiated a flurry of excitement and could be credited by inspiring the many scientists now studying applications of carbon nanotubes. Though Iijima has been given much of the credit for discovering carbon nanotubes, it turns out that the timeline of carbon nanotubes goes back much further than 1991. In 1952 L. V. Radushkevich and V. M. Lukyanovich published clear images of 50 nanometer diameter tubes made of carbon in the Soviet *Journal of Physical Chemistry*. This discovery was largely unnoticed, as the article was published in Russian, and Western scientists' access to Soviet press was limited during the Cold War. Before they came to be known as carbon nanotubes, in 1976, Morinobu Endo of CNRS observed hollow tubes of rolled up graphite sheets synthesised by a chemical vapour-growth technique. The first specimens observed would later come to be known as single-walled carbon nanotubes (SWNTs). The three scientists have been the first ones to show images of a nanotube with a solitary graphene wall.

Endo, in his early review of vapor-phase-grown carbon fibers (VPCF), also reminded us that he had observed a hollow tube, linearly extended with parallel carbon layer faces near the fiber core. This appears to be the observation of multi-walled carbon nanotubes at the center of the fiber. The mass-produced MWCNTs today are strongly related to the VPGCF developed by Endo. In fact, they call it the "Endo-process", out of respect for his early work and patents.

In 1979, John Abrahamson presented evidence of carbon nanotubes at the 14th Biennial Conference of Carbon at Pennsylvania State University. The conference paper described carbon nanotubes as carbon fibers that were produced on carbon anodes during arc discharge. A characterization of these fibers was given as well as hypotheses for their growth in a nitrogen atmosphere at low pressures.

In 1981, a group of Soviet scientists published the results of chemical and structural characterization of carbon nanoparticles produced by a thermocatalytical disproportionation of carbon monoxide. Using TEM images and XRD patterns, the authors suggested that their "carbon multi-layer tubular crystals" were formed by rolling graphene layers into cylinders. They speculated that by rolling graphene layers into a cylinder, many different arrangements of graphene hexagonal nets are possible. They suggested two possibilities of such arrangements: circular arrangement (armchair nanotube) and a spiral, helical arrangement (chiral tube).

In 1987, Howard G. Tennent of Hyperion Catalysis was issued a U.S. patent for the production of "cylindrical discrete carbon fibrils" with a "constant diameter between about 3.5 and about 70 nanometers..., length 10^2 times the diameter, and an outer region of multiple essentially continuous layers of ordered carbon atoms and a distinct inner core...."

Iijima's discovery of multi-walled carbon nanotubes in the insoluble material of arc-burned graphite rods in 1991 and Mintmire, Dunlap, and White's independent prediction that if single-walled carbon nanotubes could be made, then they would exhibit remarkable conducting properties helped create the initial buzz that is now associated with carbon nanotubes. Nanotube research accelerated greatly following the independent discoveries by Bethune at IBM and Iijima at NEC of *single-walled* carbon nanotubes and methods to specifically produce them by adding transition-metal catalysts to the carbon in an arc discharge. The arc discharge technique was well-known to produce the famed Buckminster fullerene on a preparative scale, and these results appeared to extend the run of accidental discoveries relating to fullerenes. The discovery of nanotubes remains a contentious issue. Many believe that Iijima's report in 1991 is of particular importance because it brought carbon nanotubes into the awareness of the scientific community as a whole.

Silver Nanoparticle

Silver nanoparticles are nanoparticles of silver of between 1 nm and 100 nm in size. While frequently described as being 'silver' some are composed of a large percentage of silver oxide due to their large ratio of surface-to-bulk silver atoms. Numerous shapes of nanoparticles can be constructed depending on the application at hand. Commonly used are spherical silver nanoparticles but diamond, octagonal and thin sheets are also popular.

Their extremely large surface area permits the coordination of a vast number of ligands. The properties of silver nanoparticles applicable to human treatments are under investigation in laboratory and animal studies, assessing potential efficacy, toxicity, and costs.

Synthesis

Methods

Silver nanoparticles can be synthesized by several methods. They can be divided into three broad categories: wet chemistry, ion implantation, or biogenic synthesis.

Wet Chemistry

Several wet chemical methods have been developed to form silver nanoparticles. Each of these methods use the same basic mechanism: a silver ion complex is reduced to colloidal silver in the presence of a reducing agent, which oxidizes and gives back the missing electron, so that the silver ion is reduced back to its metallic state. The newly formed silver nanoparticles can then also be stabilized with a stabilizing/capping agent, which stabilizes the energy of the surface of the newly formed nanoparticle to prevent them from aggregate to larger particles. These methods include the use of reducing sugars, citrate reduction, reduction via sodium borohydride, the silver mirror reaction, the polyol process, seed-mediated growth, and light-mediated growth. Each of these methods, or combination of methods, will offer differing degrees of control over the size distribution as well as distributions of geometric arrangements of the nanoparticle. A new, very promising wet-chemical technique was found by Elsupikhe et al. (2015). They have developed a green ultrasonically-assisted synthesis. Under ultrasound treatment, silver nanoparticles (AgNP) are synthesized with κ-carrageenan as a natural stabilizer. The reaction is performed at ambient temperature and produces silver nanoparticles with fcc crystal structure without impurities. The concentration of κ-carrageenan is used to influence particle size distribution of the AgNPs.

Reduction with Reducing Sugars

Reducing sugars includes glucose, fructose, maltose, maltodextrin, etc. but not sucrose, and is the simpliest method to reduce silver ions back to silver nanoparticles. And if the pH is over 7 the glucose molucule in the reducing sugar will "open up" its ring structure and attach to the surface of the nanoparticles, and thereby act as a weak stabilizing agent as well.

Citrate Reduction

An early, and very common, method for synthesizing silver nanoparticles is citrate reduction. Citrate reduction involves the reduction of silver nitrate, $AgNO_3$, to colloidal silver using trisodium citrate, $Na_3C_6H_5O_7$ at elevated temperature (~100 °C). In this synthesis, the citrate ion acts as both the reducing and capping agent. This method is useful due to its relative ease and short reaction time. However, the silver particles formed may exhibit broad size distributions and/or form several different particle geometries simultaneously.

Reduction Via Sodium Borohydride

A similar method is the reduction of silver nitrate using sodium borohydride, NaBH4. Sodium borohydride is a stronger reducing agent than citrate, but it is not thought to participate in surface stabilization. Thus, a separate capping molecule must be used in addition to the sodium borohydride. Examples of stabilizers used include citrate, poly(vinyl pyrrolidone) (PVP), bovine serum albumin, and cetyltrimethylammonium bromide (CTAB).

Polyol Process

The polyol process is a particularly useful method because it yields a high degree of control over both the size and geometry of the resulting nanoparticles. In general, the polyol synthesis begins with the heating of a polyol compound such as ethylene glycol, 1,5-pentanediol, or 1,2-propylene glycol7. An Ag+ species and a capping agent are added (although the polyol itself is also often the capping agent). The Ag+ species is then reduced by the polyol to colloidal nanoparticles. The polyol process is highly sensitive to reaction conditions such as temperature, chemical environment, and concentration of substrates.(add refs again here) Therefore, by changing these variables, various sizes and geometries can be selected for such as quasi-spheres, pyramids, spheres, and wires. Further study has examined the mechanism for this process as well as resulting geometries under various reaction conditions in greater detail.

Seed-mediated Growth

Seed-mediated growth is the process in which seed nanoparticles, formed from a reduction of Ag+ to colloidal silver by any method, are placed in a growth solution. The growth solution contains an additional silver species that deposits more silver atoms onto the original seeds. Several experimental factors are manipulated to achieve size and geometry control in this method such as concentration of the substrates and the capping agent used. Parameters controlling the size and geometry of the particles is described in greater detail in various studies.

Light-mediated Growth

Light-mediated syntheses have also been explored where light can promote formation of various silver nanoparticle morphologies.

Silver Mirror Reaction

The silver mirror reaction involves the conversion of silver nitrate to Ag(NH3)OH. Ag(NH3)OH is subsequently reduced into colloidal silver using an aldehyde containing molecule such as a sugar. The silver mirror reaction is as follows:

$$2(Ag(NH_3)_2)^+ + RCHO + 2OH^- RCOOH + 2Ag + 4NH_3$$

The size and shape of the nanoparticles produced are difficult to control and often have wide distributions. However, this method is often used to apply thin coatings of silver particles onto surfaces and further study into producing more uniformly sized nanoparticles is being done.

Ion Implantation

Ion implantation has been used to create silver nanoparticles embedded in glass, polyurethane, silicone, polyethylene, and poly(methyl methacrylate). Particles are embedded in the substrate by means of bombardment at high accelerating voltages. At a fixed current density of the ion beam up to a certain value, the size of the embedded silver nanoparticles has been found to be monodisperse within the population, after which only an increase in the ion concentration is observed. A further increase in the ion beam dose has been found to reduce both the nanoparticle size and density in the target substrate, whereas an ion beam operating at a high accelerating voltage with a gradually increasing current density has been found to result in a gradual increase in the nanoparticle size. There are a few competing mechanisms which may result in the decrease in nanoparticle size; destruction of NPs upon collision, sputtering of the sample surface, particle fusion upon heating and dissociation.

The formation of embedded nanoparticles is complex, and all of the controlling parameters and factors have not yet been investigated. Computer simulation is still difficult as it involves processes of diffusion and clustering, however it can be broken down into a few different sub-processes such as implantation, diffusion, and growth. Upon implantation, silver ions will reach different depths within the substrate which approaches a Gaussian distribution with the mean centered at X depth. High temperature conditions during the initial stages of implantation will increase the impurity diffusion in the substrate and as a result limit the impinging ion saturation, which is required for nanoparticle nucleation. Both the implant temperature and ion beam current density are crucial to control in order to obtain a monodisperse nanoparticle size and depth distribution. A low current density may be used to counter the thermal agitation from the ion beam and a buildup of surface charge. After implantation on the surface, the beam currents may be raised as the surface conductivity will increase. The rate at which impurities diffuse drops quickly after the formation of the nanoparticles, which act as a mobile ion trap. This suggests that the beginning of the implantation process is critical for control of the spacing and depth of the resulting nanoparticles, as well as control of the substrate temperature and ion beam density. The presence and nature of these particles can be analyzed using numerous spectroscopy and microscopy instruments. Nanoparticles synthesized in the substrate exhibit surface plasmon resonances as evidenced by characteristic absorption bands; these features undergo spectral shifts depending on the nanoparticle size and surface asperities, however the optical properties also strongly depend on the substrate material of the composite.

Biogenic Synthesis

The biological synthesis of nanoparticles has provided a means for improved techniques compared to the traditional methods that call for the use of harmful reducing agents like sodium borohydride. Many of these methods could improve their environmental footprint by replacing these relatively strong reducing agents. The problems with the chemical production of silver nanoparticles is usually involves high cost and the longevity of the particles is short lived due to aggregation. The harshness of standard chemical methods has sparked the use of using biological organisms to reduce silver ions in solution into colloidal nanoparticles.

In addition, precise control over shape and size is vital during nanoparticle synthesis since the NPs therapeutic properties are intimately dependent on such factors. Hence, the primary focus of research in biogenic synthesis is in developing methods that consistently reproduce NPs with precise properties.

Use of Fungi and Bacteria

A general representation of the synthesis and applications of biogenically synthesized silver nanoparticles using plant extract.

Bacterial and fungal synthesis of nanoparticles is practical because bacteria and fungi are easy to handle and can be modified genetically with ease. This provides a means to develop biomolecules that can synthesize AgNPs of varying shapes and sizes in high yield, which is at the forefront of current challenges in nanoparticle synthesis. Fungal strains such as Verticillium and bacterial strains such as K. pneumoniae can be used in the synthesis of silver nanoparticles. When the fungus/bacteria is added to solution, protein biomass is released into the solution. Electron donating residues such as tryptophan and tyrosine reduce silver ions in solution contributed by silver nitrate. These methods have been found to effectively create stable monodisperse nanoparticles without the use of harmful reducing agents.

A method has been found of reducing silver ions by the introduction of the fungus *Fusarium oxysporum*. The nanoparticles formed in this method have a size range between 5 and 15 nm and consist of silver hydrosol. The reduction of the silver nanoparticles is thought to come from an enzymatic process and silver nanoparticles produced are extremely stable due to interactions with proteins that are excreted by the fungi.

Bacterium found in silver mines, Pseudomonas stutzeri AG259, were able to construct silver particles in the shapes of triangles and hexagons. The size of these nanoparticles had a large range in size and some of them reached sizes larger than the usual nanoscale with a size of 200 nm. The silver nanoparticles were found in the organic matrix of the bacteria.

Lactic acid producing bacteria have been used to produce silver nanoparticles. The bacteria *Lactobacillus* spp., *Pediococcus pentosaceus, Enteroccus faeciumI*, and *Lactococcus garvieae* have been found to be able to reduce silver ions into silver nanoparticles. The production of the nanoparticles takes place in the cell from the interactions between the silver ions and the organic compounds of the cell. It was found that the bacterium *Lactobacillus fermentum* created the smallest silver nanoparticles with an average size of 11.2 nm. It was also found that this bacterium produced the nanoparticles with the smallest size distribution and the nanoparticles were found mostly on the outside of the cells. It was also found that there was an increase in the pH increased the rate of which the nanoparticles were produced and the amount of particles produced.

Use of Plants

The reduction of silver ions into silver nanoparticles has also been achieved using geranium leaves. It has been found that adding geranium leaf extract to silver nitrate solutions causes there silver ions to be quickly reduced and that the nanoparticles produced are particularly stable. The silver nanoparticles produced in solution had a size range between 16 and 40 nm.

In another study different plant leaf extracts were used to reduce silver ions. It was found that out of Camellia sinensis (green tea), pine, persimmon, ginko, magnolia, and platanus that the magnolia leaf extract was the best at creating silver nanoparticles. This method created particles with a disperse size range of 15 to 500 nm, but it was also found that the particle size could be controlled by varying the reaction temperature. The speed at which the ions were reduced by the magnolia leaf extract was comparable to those of using chemicals to reduce.

The use of plants, microbes, and fungi in the production of silver nanoparticles is leading the way to more environmentally sound production of silver nanoparticles.

Products and Functionalization

Synthetic protocols for silver nanoparticle production can be modified to produce silver nanoparticles with non-spherical geometries and also to functionalize nanoparticles with different materials, such as silica. Creating silver nanoparticles of different shapes and surface coatings allows for greater control over their size-specific properties.

Anisotropic Structures

Silver nanoparticles can be synthesized in a variety of non-spherical (anisotropic) shapes. Because silver, like other noble metals, exhibits a size and shape dependent optical effect known as localized surface plasmon resonance (LSPR) at the nanoscale, the ability to synthesize Ag nanoparticles in different shapes vastly increases the ability to tune their optical behavior. For example, the wavelength at which LSPR occurs for a nanoparticle of one morphology (e.g. a sphere) will be different if that sphere is changed into a different shape. This shape dependence allows a silver nanoparticle to experience optical enhancement at a range of different wavelengths, even by keeping the size relatively constant, just by changing its shape. The applications of this shape-exploited expansion of optical behavior range from developing more sensitive biosensors to increasing the longevity of textiles.

Triangular Nanoprisms

Triangular shaped nanoparticles are a canonical type of anisotropic morphology studied for both gold and silver.

Though many different techniques for silver nanoprism synthesis exist, several methods employ a seed-mediated approach, which involves first synthesizing small (3-5 nm diameter) silver nanoparticles that offer a template for shape-directed growth into triangular nanostructures.

The silver seeds are synthesized by mixing silver nitrate and sodium citrate in aqueous solution and then rapidly adding sodium borohydride. Additional silver nitrate is added to the seed solution at low temperature, and the prisms are grown by slowly reducing the excess silver nitrate using ascorbic acid.

With the seed-mediated approach to silver nanoprism synthesis, selectivity of one shape over another can in part be controlled by the capping ligand. Using essentially the same procedure above but changing citrate to poly (vinyl pyrrolidone) (PVP) yields cube and rod-shaped nanostructures instead of triangular nanoprisms.

In addition to the seed mediated technique, silver nanoprisms can also be synthesized using a photo-mediated approach, in which preexisting spherical silver nanoparticles are transformed into triangular nanoprisms simply by exposing the reaction mixture to high intensities of light.

Nanocubes

Silver nanocubes can be synthesized using ethylene glycol as a reducing agent and PVP as a capping agent, in a polyol synthesis reaction (vide supra). A typical synthesis using these reagents involves adding fresh silver nitrate and PVP to a solution of ethylene glycol heated at 140 °C.

This procedure can actually be modified to produce another anisotropic silver nanostructure, nanowires, by just allowing the silver nitrate solution to age before using it in the synthesis. By allowing the silver nitrate solution to age, the initial nanostructure formed during the synthesis is slightly different than that obtained with fresh silver nitrate, which influences the growth process, and therefore, the morphology of the final product.

Coating with Silica

In this method, polyvinylpyrrolidone (PVP) is dissolved in water by sonication and mixed with silver colloid particles. Active stirring ensures the PVP has adsorbed to the nanoparticle surface. Centrifuging separates the PVP coated nanoparticles which are then transferred to a solution of ethanol to be centrifuged further and placed in a solution of ammonia, ethanol and $Si(OEt_4)$ (TES). Stirring for twelve hours results in the silica shell being formed consisting of a surrounding layer of silicon oxide with an ether linkage available to add functionality. Varying the amount of TES allows for different thicknesses of shells formed. This technique is popular due to the ability to add a variety of functionality to the exposed silica surface.

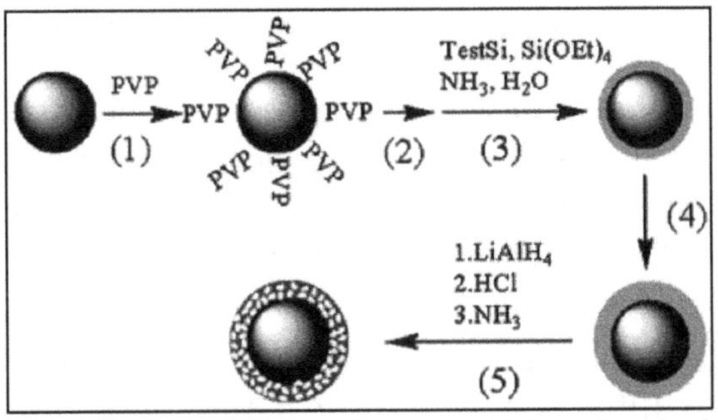

General procedure for coating colloid particles in silica. First PVP is absorbed onto the colloidal surface. These particles are put into a solution of ammonia in ethanol. the particle then begins to grow by addition of Si(OET4).

Use

Catalysis

Using silver nanoparticles for catalysis has been gaining attention in recent years. Although the most common applications are for medicinal or antibacterial purposes, silver nanoparticles have been demonstrated to show catalytic redox properties for dyes, benzene, carbon monoxide, and likely other compounds.

NOTE: This paragraph is a general description of nanoparticle properties for catalysis; it is not exclusive to silver nanoparticles. The size of a nanoparticle greatly determines the properties that it exhibits due to various quantum effects. Additionally, the chemical environment of the nanoparticle plays a large role on the catalytic properties. With this in mind, it is important to note that heterogeneous catalysis takes place by adsorption of the reactant species to the catalytic substrate. When polymers, complex ligands, or surfactants are used to prevent coalescence of the nanoparticles, the catalytic ability is frequently hindered due to reduced adsorption ability. However, these compounds can also be used in such a way that the chemical environment enhances the catalytic ability.

Supported on Silica Spheres – Reduction of Dyes

Silver nanoparticles have been synthesized on a support of inert silica spheres. The support plays virtually no role in the catalytic ability and serves as a method of preventing coalescence of the silver nanoparticles in colloidal solution. Thus, the silver nanoparticles were stabilized and it was possible to demonstrate the ability of them to serve as an electron relay for the reduction of dyes by sodium borohydride. Without the silver nanoparticle catalyst, virtually no reaction occurs between sodium borohydride and the various dyes: methylene blue, eosin, and rose Bengal.

Mesoporous Aerogel – Selective Oxidation of Benzene

Silver nanoparticles supported on aerogel are advantageous due to the higher number of active sites. The highest selectivity for oxidation of benzene to phenol was observed at low weight percent of silver in the aerogel matrix (1% Ag). This better selectivity is believed to be a result of the higher monodispersity within the aerogel matrix of the 1% Ag sample. Each weight percent solution formed different sized particles with a different width of size range.

Silver Alloy – Synergistic Oxidation of Carbon Monoxide

Au-Ag alloy nanoparticles have been shown to have a synergistic effect on the oxidation of carbon monoxide (CO). On its own, each pure-metal nanoparticle shows very poor catalytic activity for CO oxidation; together, the catalytic properties are greatly enhanced. It is proposed that the gold acts as a strong binding agent for the oxygen atom and the silver serves as a strong oxidizing catalyst, although the exact mechanism is still not completely understood. When synthesized in an Au/Ag ratio from 3:1 to 10:1, the alloyed nanoparticles showed complete conversion when 1% CO was fed in air at ambient temperature. Interestingly, the size of the alloyed particles did not play a big role in the catalytic ability. It is well known that gold nanoparticles only show catalytic properties for CO when they are ~3 nm in size, but alloyed particles up to 30 nm demonstrated excellent catalytic activity – catalytic activity better than that of gold nanoparticles on active support such as TiO2, Fe2O3, etc.

Light-enhanced

Plasmonic effects have been studied quite extensively. Until recently, there have not been studies investigating the oxidative catalytic enhancement of a nanostructure via excitation of its surface plasmon resonance. The defining feature for enhancing the oxidative catalytic ability has been identified as the ability to convert a beam of light into the form of energetic electrons that can be transferred to adsorbed molecules. The implication of such a feature is that photochemical reactions can be driven by low-intensity continuous light can be coupled with thermal energy.

The coupling of low-intensity continuous light and thermal energy has been performed with silver nanocubes. The important feature of silver nanostructures that are enabling for photocatalysis is their nature to create resonant surface plasmons from light in the visible range.

The addition of light enhancement enabled the particles to perform to the same degree as particles that were heated up to 40K greater. This is a profound finding when noting that a reduction in temperature of 25K can increase the catalyst lifetime by nearly tenfold, when comparing the photothermal and thermal process.

Biological Research

Researchers have explored the use of silver nanoparticles as carriers for delivering various payloads such as small drug molecules or large biomolecules to specific targets. Once the AgNP has had sufficient time to reach its target, release of the payload could potentially be triggered by an internal or external stimulus. The targeting and accumulation of nanoparticles may provide high payload concentrations at specific target sites and could minimize side effects.

Chemotherapy

The introduction of nanotechnology into medicine is expected to advance diagnostic cancer imaging and the standards for therapeutic drug design. Nanotechnology may uncover insight about the structure, function and organizational level of the biosystem at the nanoscale.

Silver nanoparticles can undergo coating techniques that offer a uniform functionalized surface to which substrates can be added. When the nanoparticle is coated, for example, in silica the surface exists as silicic acid. Substrates can thus be added through stable ether and ester linkages that are not degraded immediately by natural metabolic enzymes. Recent chemotherapeutic applications have designed anti cancer drugs with a photo cleavable linker, such as an ortho-nitrobenzyl bridge, attaching it to the substrate on the nanoparticle surface. The low toxicity nanoparticle complex can remain viable under metabolic attack for the time necessary to be distributed throughout the bodies systems. If a cancerous tumor is being targeted for treatment, ultraviolet light can be introduced over the tumor region. The electromagnetic energy of the light causes the photo responsive linker to break between the drug and the nanoparticle substrate. The drug is now cleaved and released in an unaltered active form to act on the cancerous tumor cells. Advantages anticipated for this method is that the drug is transported without highly toxic compounds, the drug is released without harmful radiation or relying on a specific chemical reaction to occur and the drug can be selectively released at a target tissue.

A second approach is to attach a chemotherapeutic drug directly to the functionalized surface of the silver nanoparticle combined with a nucelophilic species to undergo a displacement reaction. For example, once the nanoparticle drug complex enters or is in the vicinity of the target tissue or cells, a glutathione monoester can be administered to the site. The nucleophilic ester oxygen will attach to the functionalized surface of the nanoparticle through a new ester linkage while the drug is released to its surroundings. The drug is now active and can exert its biological function on the cells immediate to its surroundings limiting non-desirable interactions with other tissues.

Multiple Drug Resistance

A major cause for the ineffectiveness of current chemotherapy treatments is multiple drug resistance which can arise from several mechanisms.

Nanoparticles can provide a means to overcome MDR. In general, when using a targeting agent to deliver nanocarriers to cancer cells, it is imperative that the agent binds with high selectivity to molecules that are uniquely expressed on the cell surface. Hence NPs can be designed with proteins that specifically detect drug resistant cells with overexpressed transporter proteins on their surface. A pitfall of the commonly used nano-drug delivery systems is that free drugs that are released from the nanocarriers into the cytosol get exposed to the MDR transporters once again, and are exported. To solve this, 8 nm nano crystalline silver particles were modified by the addition of trans-activating transcriptional activator (TAT), derived from the HIV-1 virus, which acts as a cell penetrating peptide (CPP). Generally, AgNP effectiveness is limited due to the lack of efficient cellular uptake; however, CPP-modification has become one of the most efficient methods for improving intracellular delivery of nanoparticles. Once ingested, the export of the AgNP is prevented based on a size exclusion. The concept is simple: the nanoparticles are too large to be effluxed by the MDR transporters, because the efflux function is strictly subjected to the size of its substrates, which is generally limited to a range of 300-2000 Da. Thereby the nanoparticulates remain insusceptible to the efflux, providing a means to accumulate in high concentrations.

Antimicrobial

Introduction of silver into bacterial cells induces a high degree of structural and morphological changes, which can lead to cell death. As the silver nano particles come in contact with the bacteria, they adhere to the cell wall and cell membrane. Once bound, some of the silver passes through to the inside, and interacts with phosphate-containing compounds like DNA and RNA, while another portion adheres to the sulphur-containing proteins on the membrane. The silver-sulphur interactions at the membrane cause the cell wall to undergo structural changes, like the formation of pits and pores. Through these pores, cellular components are released into the extracellular fluid, simply due to the osmotic difference. Within the cell, the integration of silver creates a low molecular weight region where the DNA then condenses. Having DNA in a condensed state inhibits the cell's replication proteins contact with the DNA. Thus the introduction of silver nanoparticles inhibits replication and is sufficient to cause the death of the cell. Further increasing their effect, when silver comes in contact with fluids, it tends to ionize which increases the nanoparticles bactericidal activity. This has been correlated to the suppression of enzymes and inhibited expression of proteins that relate to the cell's ability to produce ATP.

Although it varies for every type of cell proposed, as their cell membrane composition varies greatly, It has been seen that in general, silver nano particles with an average size of 10 nm or less show electronic effects that greatly increase their bactericidal activity. This could also be partly due to the fact that as particle size decreases, reactivity increases due to the surface area to volume ratio increasing.

It has been noted that the introduction of silver nano particles has shown to have synergistic activity with common antibiotics already used today, such as; penicillin G, ampicillin, erythromycin, clindamycin, and vancomycin against E. coli and S. aureus. In medical equipment, it has been shown that silver nano particles drastically lower the bacterial count on devices used. However, the problem arises when the procedure is over and a new one must be done. In the process of washing the instruments a large portion of the silver nano particles become less effective due to the loss of silver ions. They are more commonly used in skin grafts for burn victims as the silver nano particles embedded with the graft provide better antimicrobial activity and result in significantly less scarring of the victim. They also show promising application as water treatment method to form clean potable water.

Other

Samsung has created and marketed a material called Silver Nano, that includes silver nanoparticles on the surfaces of household appliances.

Safety

Although silver nanoparticles are widely used in a variety of commercial products, there has only recently been a major effort to study their effects on human health. There have been several studies that describe the *in vitro* toxicity of silver nanoparticles to a variety of different organs, including the lung, liver, skin, brain, and reproductive organs. The mechanism of the toxicity of silver nanoparticles to human cells appears to be derived from oxidative stress and inflammation that is caused by the generation of reactive oxygen species (ROS) stimulated by either the Ag NPs, Ag ions, or both. For example, Park *et al.* showed that exposure of a mouse peritoneal macrophage cell line (RAW267.7) to silver nanoparticles decreased the cell viability in a concentration- and time-dependent manner. They further showed that the intracellular reduced glutathionine (GSH), which is a ROS scavenger, decreased to 81.4% of the control group of silver nanoparticles at 1.6 ppm.

Modes of Toxicity

Since silver nanoparticles undergo dissolution releasing silver ions, which is well-documented to have toxic effects, there have been several studies that have been conducted to determine whether the toxicity of silver nanoparticles is derived from the release of silver ions or from the nanoparticle itself. Several studies suggest that the toxicity of silver nanoparticles is attributed to their release of silver ions in cells as both silver nanoparticles and silver ions have been reported to have similar cytotoxicity. For example, In some cases it is reported that silver nanoparticles facilitate the release of toxic free silver ions in cells via a "Trojan-horse type mechanism," where the particle enters cells and is then ionized within the cell. However, there have been reports that suggest that a combination of silver nanoparticles and ions is responsible for the toxic effect of silver nanoparticles. Navarro *et al.* using cysteine ligands as a tool to measure the concentration of free silver in solution, determined that although initially silver ions were 18 times more likely to inhibit the photosynthesis of an algae, Chlamydomanas reinhardtii, but after 2 hours of incubation it was revealed that the algae containing silver nanoparticles were more toxic than just silver ions alone. Furthermore, there are studies that suggest that silver nanoparticles induce toxicity independent of free silver ions. For example, Asharani *et al.* compared phenotypic defects observed in zebrafish treated with silver nanoparticles and silver ions and determined that the phenotypic defects observed with silver nanoparticle treatment was not observed with silver ion-treated embryos, suggesting that the toxicity of silver nanoparticles are independent of silver ions.

Protein channels and nuclear membrane pores can often be in the size range of 9 nm to 10 nm in diameter. Small silver nanoparticles constructed of this size have the ability to not only pass through the membrane to interact with internal structures but also to be become lodged within the membrane. Silver nanoparticle depositions in the membrane can impact regulation of solutes, exchange of proteins and cell recognition. Exposure to silver nanoparticles has been associated with "inflammatory, oxidative, genotoxic, and cytotoxic consequences"; the silver particulates primarily accumulate in the liver. but have also been shown to be toxic in other organs including the brain. Nano-silver applied to tissue-cultured human cells leads to the formation of free radicals, raising concerns of potential health risks.

- Allergic reaction: There have been several studies conducted that show a precedence for allerginicity of silver nanoparticles.

- Argyria and staining: Ingested silver or silver compounds, including colloidal silver, can cause a condition called argyria, a discoloration of the skin and organs.In 2006, there was a case study of a 17-year-old man, who sustained burns to 30% of his body, and experienced a temporary bluish-grey hue after several days of treatment with Acticoat, a brand of wound dressing containing silver nanoparticles. Argyria is the deposition of silver in deep tissues, a condition that cannot happen on a temporary basis, raising the question of whether the cause of the man's discoloration was argyria or even a result of the silver treatment. Silver dressings are known to cause a "transient discoloration" that dissipates in 2–14 days, but not a permanent discoloration.

- Silzone heart valve: St. Jude Medical released a mechanical heart valve with a silver coated sewing cuff (coated using ion beam-assisted deposition) in 1997. The valve was designed to reduce the instances of endocarditis. The valve was approved for sale in Canada, Europe, the United States, and most other markets around the world. In a post-commercialization study, researchers showed that the valve prevented tissue ingrowth, created paravalvular leakage, valve loosening, and in the worst cases explantation. After 3 years on the market and 36,000 implants, St. Jude discontinued and voluntarily recalled the valve.

Hybrid Material

Hybrid materials are composites consisting of two constituents at the nanometer or molecular level. Commonly one of these compounds is inorganic and the other one organic in nature. Thus, they differ from traditional composites where the constituents are at the macroscopic (micrometer to millimeter) level. Mixing at the microscopic scale leads to a more homogeneous material that either show characteristics in between the two original phases or even new properties.

Introduction

Hybrid Materials in Nature

Many natural materials consist of inorganic and organic building blocks distributed on the nanoscale. In most cases the inorganic part provides mechanical strength and an overall structure to the natural objects while the organic part delivers bonding between the inorganic building blocks and/or the soft tissue. Typical examples of such materials are bone, or nacre.

Development of Hybrid Materials

The first hybrid materials were the paints made from inorganic and organic components that were used thousands of years ago. Rubber is an example of the use of inorganic materials as fillers for organic polymers. The sol–gel process developed in the 1930s was one of the major driving forces what has become the broad field of inorganic–organic hybrid materials.

Classification

Hybrid materials can be classified based on the possible interactions connecting the inorganic and organic species. *Class I* hybrid materials are those that show weak interactions between the two phases, such as van der Waals, hydrogen bonding or weak electrostatic interactions. *Class II* hybrid materials are those that show strong chemical interactions between the components such as covalent bonds.

Structural properties can also be used to distinguish between various hybrid materials. An organic moiety containing a functional group that allows the attachment to an inorganic network, e.g. a trialkoxysilane group, can act as a *network modifier* because in the final structure the inorganic network is only modified by the organic group. Phenyltrialkoxysilanes are an example for such compounds; they modify the silica network in the sol–gel process via the reaction of the trialkoxysilane group without supplying additional functional groups intended to undergo further chemical reactions to the material formed. If a reactive functional group is incorporated the system is called a *network functionalizer*. The situation is different if two or three of such anchor groups modify an organic segment; this leads to materials in which the inorganic group is afterwards an integral part of the hybrid network. The latter type of system is known as *network builder*.

Blends are formed if no strong chemical interactions exist between the inorganic and organic building blocks. One example for such a material is the combination of inorganic clusters or particles with organic polymers lacking a strong (e.g. covalent) interaction between the components. In this case a material is formed that consists for example of an organic polymer with entrapped discrete inorganic moieties in which, depending on the functionalities of the components, for example weak crosslinking occurs by the entrapped inorganic units through physical interactions or the inorganic components are entrapped in a crosslinked polymer matrix. If an inorganic and an organic network interpenetrate each other without strong chemical interactions, so called interpenetrating networks (IPNs) are formed, which is for example the case if a sol–gel material is formed in presence of an organic polymer or vice versa. Both materials described belong to class I hybrids. Class II hybrids are formed when the discrete inorganic building blocks, e.g. clusters, are covalently bonded to the organic polymers or inorganic and organic polymers are covalently connected with each other.

Distinction between Nanocomposites and Hybrid Materials

The term nanocomposite is used if one of the structural units, either the organic or the inorganic, is in a defined size range of 1–100 nm. Large molecular building blocks for hybrid materials, such as large inorganic clusters, may be of the nanometer length scale. The term nanocomposites is generally used if discrete structural units in the respective size regime are used and the term hybrid materials is more often used if the inorganic units are formed in situ by molecular precursors, for example applying sol–gel reactions. Examples of discrete inorganic units for nanocomposites are nanoparticles, nanorods and carbon nanotubes. The biggest distinction between a nanocomposite and a hybrid is that a hybrid material possesses a property that does not exist in either of the parent components.<G. L. Drisko, C. Sanchez, Eur. J. Inorg. Chem. 2012, 2012, 5097>

Advantages of Hybrid Materials over Traditional Composites

- Inorganic clusters or nanoparticles with specific optical, electronic or magnetic properties can be incorporated in organic polymer matrices.
- Contrary to pure solid state inorganic materials that often require a high temperature treatment for their processing, hybrid materials show a more polymer-like handling, either because of their large organic content or because of the formation of crosslinked inorganic networks from small molecular precursors just like in polymerization reactions.
- Light scattering in homogeneous hybrid material can be avoided and therefore optical transparency of the resulting hybrid materials and nanocomposites can be achieved.

Synthesis

Two different approaches can be used for the formation of hybrid materials: Either well-defined preformed building blocks are applied that react with each other to form the final hybrid material in which the precursors still at least partially keep their original integrity or one or both structural units are formed from the precursors that are transformed into a new (network) structure. It is important that the interface between the inorganic and organic materials which has to be tailored to overcome serious problems in the preparation of hybrid materials. Different building blocks and approaches can be used for their preparation and these have to be adapted to bridge the differences of inorganic and organic materials.

Building Block Approach

Building blocks at least partially keep their molecular integrity throughout the material formation, which means that structural units that are present in these sources for materials formation can also be found in the final material. At the same time typical properties of these building blocks usually survive the matrix formation, which is not the case if material precursors are transferred into novel materials. Representative examples of such well-defined building blocks are modified inorganic clusters or nanoparticles with attached reactive organic groups.

Cluster compounds often consist of at least one functional group that allows an interaction with an organic matrix, for example by copolymerization. Depending on the number of groups that can interact, these building blocks are able to modify an organic matrix (one functional group) or form partially or fully crosslinked materials (more than one group). For instance, two reactive groups can lead to the formation of chain structures. If the building blocks contain at least three reactive groups they can be used without additional molecules for the formation of a crosslinked material.

Beside the molecular building blocks mentioned, nanosized building blocks, such as particles or nanorods, can also be used to form nanocomposites. The building block approach has one large advantage compared with the in situ formation of the inorganic or organic entities: because at least one structural unit (the building block) is well-defined and usually does not undergo significant structural changes during the matrix formation, better structure–property predictions are possible. Furthermore, the building blocks can be designed in such a way to give the best performance in the materials' formation, for example good solubility of inorganic compounds in organic monomers by surface groups showing a similar polarity as the monomers.

In recent years many building blocks have been synthesized and used for the preparation of hybrid materials. Chemists can design these compounds on a molecular scale with highly sophisticated methods and the resulting systems are used for the formation of functional hybrid materials. Many future applications, in particular in nanotechnology, focus on a bottom-up approach in which complex structures are hierarchically formed by these small building blocks. This idea is also one of the driving forces of the building block approach in hybrid materials.

In Situ Formation of The Components

The in situ formation of the hybrid materials is based on the chemical transformation of the precursors used throughout materials' preparation. Typically this is the case if organic polymers are formed but also if the sol–gel process is applied to produce the inorganic component. In these cases well-defined discrete molecules are transformed to multidimensional structures, which often show totally different properties from the original precursors. Generally simple, commercially available molecules are applied and the internal structure of the final material is determined by the composition of these precursors but also by the reaction conditions. Therefore control over the latter is a crucial step in this process. Changing one parameter can often lead to two very different materials. If, for example, the inorganic species is a silica derivative formed by the sol–gel process, the change from base to acid catalysis makes a large difference because base catalysis leads to a more particle-like microstructure while acid catalysis leads to a polymer-like microstructure. Hence, the final performance of the derived materials is strongly dependent on their processing and its optimization.

In Situ Formation of Inorganic Materials

Many of the classical inorganic solid state materials are formed using solid precursors and high temperature processes, which are often not compatible with the presence of organic groups because they are decomposed at elevated temperatures. Hence, these high temperature processes are not suitable for the in situ formation of hybrid materials. Reactions that are employed should have more the character of classical covalent bond formation in solutions. One of the most prominent processes which fulfill these demands is the sol–gel process. However, such rather low temperature processes often do not lead to the thermodynamically most stable structure but to kinetic products, which has some implications for the structures obtained. For example low temperature derived inorganic materials are often amorphous or crystallinity is only observed on a very small length scale, i.e. the nanometer range. An example of the latter is the formation of metal nanoparticles in organic or inorganic matrices by reduction of metal salts or organometallic precursors.

Some methods of in situ formation of inorganic materials are:

- Sol-gel process.
- Nonhydrolytic sol–gel process.
- Sol–gel reactions of non-silicates.

Formation of Organic Polymers in Presence of Preformed Inorganic Materials

If the organic polymerization occurs in the presence of an inorganic material to form the hybrid material one has to distinguish between several possibilities to overcome the incompatibility of the two species. The inorganic material can either have no surface functionalization but the bare material surface; it can be modified with nonreactive organic groups (e.g. alkyl chains); or it can contain reactive surface groups such as polymerizable functionalities. Depending on these prerequisites the material can be pretreated, for example a pure inorganic surface can be treated with surfactants or silane coupling agents to make it compatible with the organic monomers, or functional monomers can be added that react with the surface of the inorganic material. If the inorganic component has nonreactive organic groups attached to its surface and it can be dissolved in a monomer which is subsequently polymerized, the resulting material after the organic polymerization, is a blend. In this case the inorganic component interact only weakly or not at all with the organic polymer; hence, a class I material is formed. Homogeneous materials are only obtained in this case if agglomeration of the inorganic components in the organic environment is prevented. This can be achieved if the interactions between the inorganic components and the monomers are better or at least the same as between the inorganic components. However, if no strong chemical interactions are formed, the long-term stability of a once homogeneous material is questionable because of diffusion effects in the resulting hybrid material. The stronger the respective interaction between the components, the more stable is the final material. The strongest interaction is achieved if class II materials are formed, for example with covalent interactions.

Hybrid Materials by Simultaneous Formation of both Components

Simultaneous formation of the inorganic and organic polymers can result in the most homogeneous type of interpenetrating networks. Usually the precursors for the sol–gel process are mixed with monomers for the organic polymerization and both processes are carried out at the same time with or without solvent. Applying this method, three processes are competing with each other:

- The kinetics of the hydrolysis and condensation forming the inorganic phase.
- The kinetics of the polymerization of the organic phase.
- The thermodynamics of the phase separation between the two phases.

By tailoring the kinetics of the two polymerizations in such a way that they occur simultaneously and rapidly enough, phase separation is avoided or minimized. Additional parameters such as attractive interactions between the two moieties, as described above can also be used to avoid phase separation.

One problem that also arises from the simultaneous formation of both networks is the sensitivity of many organic polymerization processes for sol–gel conditions or the composition of the materials formed. Ionic polymerizations, for example, often interact with the precursors or intermediates formed in the sol–gel process. Therefore, they are not usually applied in these reactions.

Applications

- Decorative coatings obtained by the embedding of organic dyes in hybrid coatings.
- Scratch-resistant coatings with hydrophobic or anti-fogging properties.
- Nanocomposite based devices for electronic and optoelectronic applications including light-emitting diodes, photodiodes, solar cells, gas sensors and field effect transistors.
- Fire retardant materials for construction industry.
- Nanocomposite based dental filling materials.
- Composite electrolyte materials for applications such as solid-state lithium batteries or supercapacitors.
- Proton conducting membranes used in fuel cells.
- Antistatic / anti-reflection coatings.
- Corrosion protection.
- Porous hybrid materials.

Nanowire

A nanowire is a nanostructure, with the diameter of the order of a nanometer (10^{-9} meters). It can also be defined as the ratio of the length to width being greater than 1000. Alternatively, nanowires can be defined as structures that have a thickness or diameter constrained to tens of nanometers or less and an unconstrained length. At these scales, quantum mechanical effects are important – which coined the term "quantum wires". Many different types of nanowires exist, including superconducting (e.g., YBCO), metallic (e.g., Ni, Pt, Au), semiconducting (e.g., Si, InP, GaN, etc.), and insulating (e.g., SiO2, TiO2). Molecular nanowires are composed of repeating molecular units either organic (e.g. DNA) or inorganic (e.g. $Mo_6S_{9-x}I_x$).

Overview

Typical nanowires exhibit aspect ratios (length-to-width ratio) of 1000 or more. As such they are often referred to as one-dimensional (1-D) materials. Nanowires have many interesting properties that are not seen in bulk or 3-D (three-dimensional) materials. This is because electrons in nanowires are quantum confined laterally and thus occupy energy levels that are different from the traditional continuum of energy levels or bands found in bulk materials.

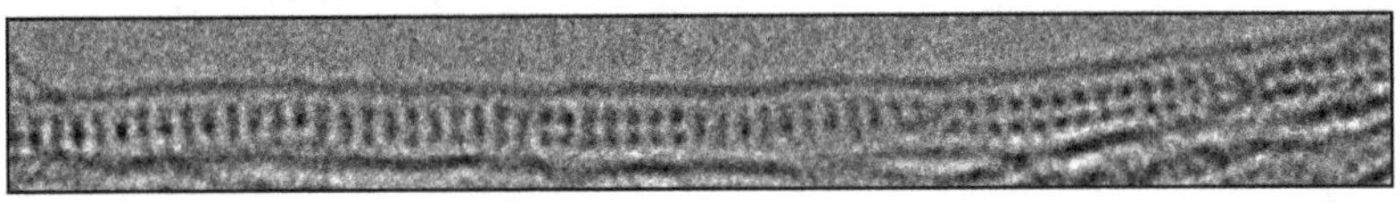

Crystalline 2×2-atom tin selenide nanowire grown inside a single-wall carbon nanotube (tube diameter ~1 nm).

Peculiar features of this quantum confinement exhibited by certain nanowires manifest themselves in discrete values of the electrical conductance. Such discrete values arise from a quantum mechanical restraint on the number of electrons that can travel through the wire at the nanometer scale. These discrete values are often referred to as the quantum of conductance and are integer multiples of

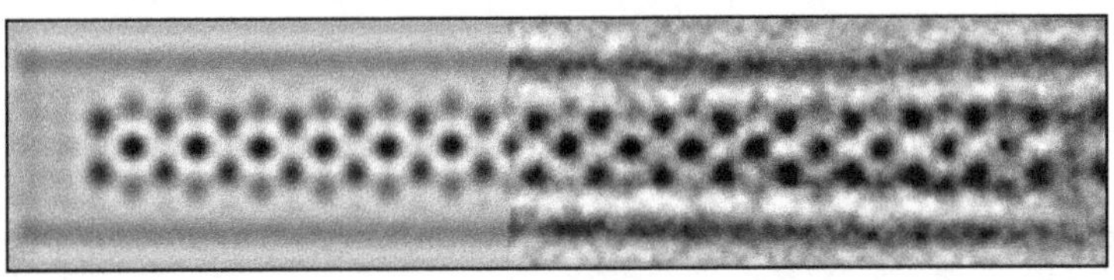

A noise-filtered HRTEM image of a HgTe extreme nanowire embedded down the central pore of a SWCNT. The image is also accompanied by a simulation of the crystal structure.

They are inverse of the well-known resistance unit h/e^2, which is roughly equal to 25812.8 ohms, and referred to as the von Klitzing constant R_K (after Klaus von Klitzing, the discoverer of exact quantization). Since 1990, a fixed conventional value $R_{K\text{-}90}$ is accepted.

Examples of nanowires include inorganic molecular nanowires ($Mo_6S_{9-x}I_x$, $Li_2Mo_6Se_6$), which can have a diameter of 0.9 nm and be hundreds of micrometers long. Other important examples are based on semiconductors such as InP, Si, GaN, etc., dielectrics (e.g. SiO_2,TiO_2), or metals (e.g. Ni, Pt).

There are many applications where nanowires may become important in electronic, opto-electronic and nanoelectromechanical devices, as additives in advanced composites, for metallic interconnects in nanoscale quantum devices, as field-emitters and as leads for biomolecular nanosensors.

Synthesis of Nanowires

There are two basic approaches to synthesizing nanowires: top-down and bottom-up. A top-down approach reduces a large piece of material to small pieces, by various means such as lithography or electrophoresis. A bottom-up approach synthesizes the nanowire by combining constituent adatoms. Most synthesis techniques use a bottom-up approach.

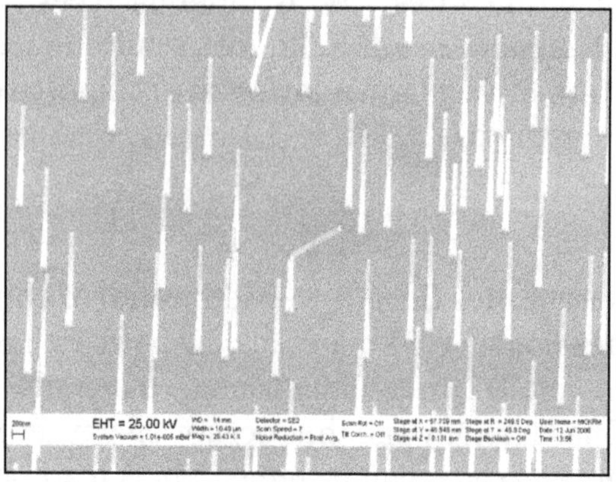

An SEM image of epitaxial nanowire heterostructures grown from catalytic gold nanoparticles.

Nanowire production uses several common laboratory techniques, including suspension, electrochemical deposition, vapor deposition, and VLS growth. Ion track technology enables growing homogeneous and segmented nanowires down to 8 nm diameter.

Suspension

A suspended nanowire is a wire produced in a high-vacuum chamber held at the longitudinal extremities. Suspended nanowires can be produced by:

- The chemical etching of a larger wire
- The bombardment of a larger wire, typically with highly energetic ions
- Indenting the tip of a STM in the surface of a metal near its melting point, and then retracting it

VLS Growth

A common technique for creating a nanowire is vapor-liquid-solid method (VLS). This process can produce crystalline nanowires of some semiconductor materials. It uses a source material from either laser ablated particles or a feed gas such as silane.

VLS synthesis requires a catalyst. For nanowires, the best catalysts are liquid metal (such as gold) nanoclusters, which can either be self-assembled from a thin film by dewetting, or purchased in colloidal form and deposited on a substrate.

The source enters these nanoclusters and begins to saturate them. On reaching supersaturation, the source solidifies and grows outward from the nanocluster. Simply turning off the source can adjust the final length of the nanowire. Switching sources while still in the growth phase can create compound nanowires with super-lattices of alternating materials.

A single-step vapour phase reaction at elevated temperature synthesises inorganic nanowires such as $Mo_6S_{9-x}I_x$. From another point of view, such nanowires are cluster polymers.

Solution-phase Synthesis

Solution-phase synthesis refers to techniques that grow nanowires in solution. They can produce nanowires of many types of materials. Solution-phase synthesis has the advantage that it can produce very large quantities, compared to other methods. In one technique, the polyol synthesis, ethylene glycol is both solvent and reducing agent. This technique is particularly versatile at producing nanowires of lead, platinum, and silver.

The supercritical fluid-liquid-solid growth method can be used to synthesize semiconductor nanowires, e.g., Si and Ge. By using metal nanocrystals as seeds, Si and Ge organometallic precursors are fed into a reactor filled with a supercritical organic solvent, such as toluene. Thermolysis results in degradation of the precursor, allowing release of Si or Ge, and dissolution into the metal nanocrystals. As more of the semiconductor solute is added from the supercritical phase (due to a concentration gradient), a solid crystallite precipitates, and a nanowire grows uniaxially from the nanocrystal seed.

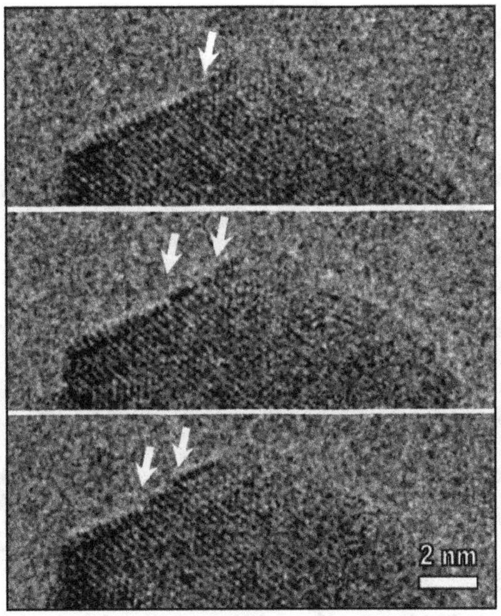

In situ observation of CuO nanowire growth.

Non-catalytic Growth

Nanowires can be also grown without the help of catalysts, which gives an advantage of pure nanowires and minimizes the number of technological steps. The simplest methods to obtain metal oxide nanowires use ordinary heating of the metals, e.g. metal wire heated with battery, by Joule heating in air can be easily done at home. The vast majority of nanowire-formation mechanisms are explained through the use of catalytic nanoparticles, which drive the nanowire growth and are either added intentionally or generated during the growth. However the mechanisms for catalyst-free growth of nanowires (or whiskers) were known from 1950s. Spontaneous nanowire formation by non-catalytic methods were explained by the dislocation present in specific directions or the growth anisotropy of various crystal faces. More recently, after microscopy advancement, the nanowire growth driven by screw dislocations or twin boundaries were demonstrated. The picture on the right shows a single atomic layer growth on the tip of CuO nanowire, observed by in situ TEM microscopy during the non-catalytic synthesis of nanowire.

Physics of Nanowires

Conductivity of Nanowires

Several physical reasons predict that the conductivity of a nanowire will be much less than that of the corresponding bulk material. First, there is scattering from the wire boundaries, whose effect will be very significant whenever the wire width is below the free electron mean free path of the bulk material. In copper, for example, the mean free path is 40 nm. Copper nanowires less than 40 nm wide will shorten the mean free path to the wire width.

Nanowires also show other peculiar electrical properties due to their size. Unlike single wall carbon nanotubes, whose motion of electrons can fall under the regime of ballistic transport (meaning the electrons can travel freely from one electrode to the other), nanowire conductivity is strongly influenced by edge effects. The edge effects come from atoms that lay at the nanowire surface and are not fully bonded to neighboring atoms like the atoms within the bulk of the nanowire. The unbonded atoms are often a source of defects within the nanowire, and may cause the nanowire to conduct electricity more poorly than the bulk material. As a nanowire shrinks in size, the surface atoms become more numerous compared to the atoms within the nanowire, and edge effects become more important.

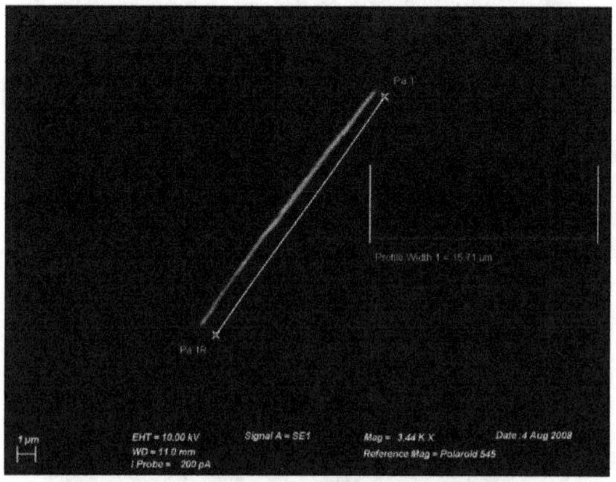

An SEM image of a 15 micrometer nickel wire.

Furthermore, the conductivity can undergo a quantization in energy: i.e. the energy of the electrons going through a nanowire can assume only discrete values, which are multiples of the conductance quantum $G = 2e^2/h$ (where e is the charge of the electron and h is the Planck constant.

The conductivity is hence described as the sum of the transport by separate *channels* of different quantized energy levels. The thinner the wire is, the smaller the number of channels available to the transport of electrons.

This quantization has been demonstrated by measuring the conductivity of a nanowire suspended between two electrodes while pulling it: as its diameter reduces, its conductivity decreases in a stepwise fashion and the plateaus correspond to multiples of G.

The quantization of conductivity is more pronounced in semiconductors like Si or GaAs than in metals, due to their lower electron density and lower effective mass. It can be observed in 25 nm wide silicon fins, and results in increased threshold voltage. In practical terms, this means that a MOSFET with such nanoscale silicon fins, when used in digital applications, will need a higher gate (control) voltage to switch the transistor on.

Welding Nanowires

To incorporate nanowire technology into industrial applications, researchers in 2008 developed a method of welding nanowires together: a sacrificial metal nanowire is placed adjacent to the ends of the pieces to be joined (using the manipulators of a scanning electron microscope); then an electric current is applied, which fuses the wire ends. The technique fuses wires as small as 10 nm.

For nanowires with diameters less than 10 nm, existing welding techniques, which require precise control of the heating mechanism and which may introduce the possibility of damage, will not be practical. Recently scientists discovered that single-crystalline ultrathin gold nanowires with diameters ~3–10 nm can be "cold-welded" together within seconds by mechanical contact alone, and under remarkably low applied pressures (unlike macro- and micro-scale cold welding process). High-resolution transmission electron microscopy and in situ measurements reveal that the welds are nearly perfect, with the same crystal orientation, strength and electrical conductivity as the rest of the nanowire. The high quality of the welds is attributed to the nanoscale sample dimensions, oriented-attachment mechanisms and mechanically assisted fast surface diffusion. Nanowire welds were also demonstrated between gold and silver, and silver nanowires (with diameters ~5–15 nm) at near room temperature, indicating that this technique may be generally applicable for ultrathin metallic nanowires. Combined with other nano- and microfabrication technologies, cold welding is anticipated to have potential applications in the future bottom-up assembly of metallic one-dimensional nanostructures.

Applications of Nanowires

Electronic Devices

Nanowires can be used for transistors. Transistors are used widely as fundamental building element in today's electronic circuits. As predicted by Moore's law, the dimension of transistors is shrinking smaller and smaller into nanoscale. One of the key challenges of building future nanoscale transistors is ensuring good gate control over the channel. Due to the high aspect ratio, if the gate dielectric is wrapped around the nanowire channel, we can get good control of channel electrostatic potential, thereby turning the transistor on and off efficiently.

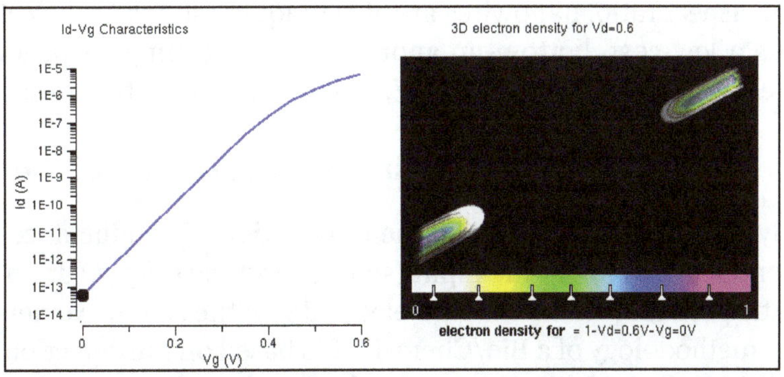

Atomistic simulation result for formation of inversion channel (electron density) and attainment of threshold voltage (IV) in a nanowire MOSFET. Note that the threshold voltage for this device lies around 0.45V.

To create active electronic elements, the first key step was to chemically dope a semiconductor nanowire. This has already been done to individual nanowires to create p-type and n-type semiconductors.

The next step was to find a way to create a p–n junction, one of the simplest electronic devices. This was achieved in two ways. The first way was to physically cross a p-type wire over an n-type wire. The second method involved chemically doping a single wire with different dopants along the length. This method created a p-n junction with only one wire.

After p-n junctions were built with nanowires, the next logical step was to build logic gates. By connecting several p-n junctions together, researchers have been able to create the basis of all logic circuits: the AND, OR, and NOT gates have all been built from semiconductor nanowire crossings.

In August 2012, researchers reported constructing the first NAND gate from undoped silicon nanowires. This avoids the problem of how to achieve precision doping of complementary nanocircuits, which is unsolved. They were able to control the Schottky barrier to achieve low-resistance contacts by placing a silicide layer in the metal-silicon interface.

It is possible that semiconductor nanowire crossings will be important to the future of digital computing. Though there are other uses for nanowires beyond these, the only ones that actually take advantage of physics in the nanometer regime are electronic.

In addition, nanowires are also being studied for use as photon ballistic waveguides as interconnects in quantum dot/quantum effect well photon logic arrays. Photons travel inside the tube, electrons travel on the outside shell.

When two nanowires acting as photon waveguides cross each other the juncture acts as a quantum dot.

Conducting nanowires offer the possibility of connecting molecular-scale entities in a molecular computer. Dispersions of conducting nanowires in different polymers are being investigated for use as transparent electrodes for flexible flat-screen displays.

Because of their high Young's moduli, their use in mechanically enhancing composites is being investigated. Because nanowires appear in bundles, they may be used as tribological additives to improve friction characteristics and reliability of electronic transducers and actuators.

Because of their high aspect ratio, nanowires are also uniquely suited to dielectrophoretic manipulation, which offers a low-cost, bottom-up approach to integrating suspended dielectric metal oxide nanowires in electronic devices such as UV, water vapor, and ethanol sensors.

Sensing of Proteins and Chemicals Using Semiconductor Nanowires

In an analogous way to FET devices in which the modulation of conductance (flow of electrons/holes) in the semiconductor, between the input (source) and the output (drain) terminals, is controlled by electrostatic potential variation (gate-electrode) of the charge carriers in the device conduction channel, the methodology of a Bio/Chem-FET is based on the detection of the local change in charge density, or so-called "field effect", that characterizes the recognition event between a target molecule and the surface receptor.

This change in the surface potential influences the Chem-FET device exactly as a 'gate' voltage does, leading to a detectable and measurable change in the device conduction. When these devices are fabricated using semiconductor nanowires as the transistor element the binding of a chemical or biological species to the surface of the sensor can lead to the depletion or accumulation of charge carriers in the "bulk" of the nanometer diameter nanowire i.e. (small cross section available for conduction channels). Moreover, the wire, which serves as a tunable conducting channel, is in close contact with the sensing environment of the target, leading to a short response time, along with orders of magnitude increase in the sensitivity of the device as a result of the huge S/V ratio of the nanowires.

While several inorganic semiconducting materials such as Si, Ge, or metal oxides (e.g. In2O3, SnO2, ZnO, etc.) have been used for the preparation of nanowires. Silicon nanowires are usually the material of choice when fabricating nanowire FET-based chemo/biosensors.

Several examples of the use of silicon nanowire sensing devices include the ultra sensitive, real-time sensing of biomarker proteins for cancer, detection of single virus particles, and the detection of nitro-aromatic explosive materials such as 2,4,6 Tri-nitrotoluene (TNT) in sensitives superior to these of canines. Silicon nanowires could also be used in their twisted form, as electromechanical devices, to measure intermolecular forces with great precision.

Limitations of Sensing With Silicon Nanowire FET Devices

Generally, the charges on dissolved molecules and macromolecules are screened by dissolved counterions, since in most cases molecules bound to the devices are separated from the sensor surface by approximately 2–12 nm (the size of the receptor proteins or DNA linkers bound to the sensor surface). As a result of the screening, the electrostatic potential that arises from charges on the analyte molecule decays exponentially toward zero with distance. Thus, for optimal sensing, the Debye length must be carefully selected for nanowire FET measurements. One approach of overcoming this limitation employs fragmentation of the antibody-capturing units and control over surface receptor density, allowing more intimate binding to the nanowire of the target protein. This approach proved useful for dramatically enhancing the sensitivity of cardiac biomarkers (e.g. Troponin) detection directly from serum for the diagnosis of acute myocardial infarction.

Nanocellulose

Nanocellulose is a term referring to nano-structured cellulose. This may be either cellulose nanofibers (CNF) also called microfibrillated cellulose (MFC), nanocrystalline cellulose (NCC), or bacterial nanocellulose, which refers to nano-structured cellulose produced by bacteria.

CNF is a material composed of nanosized cellulose fibrils with a high aspect ratio (length to width ratio). Typical lateral dimensions are 5–20 nanometers and longitudinal dimension is in a wide range, typically several micrometers. It is pseudo-plastic and exhibits the property of certain gels or fluids that are thick (viscous) under normal conditions, but flow (become thin, less viscous) over time when shaken, agitated, or otherwise stressed. This property is known as thixotropy. When the shearing forces are removed the gel regains much of its original state. The fibrils are isolated from any cellulose containing source including wood-based fibers (pulp fibers) through high-pressure, high temperature and high velocity impact homogenization, grinding or microfluidization.

Nanocellulose.

Nanocellulose can also be obtained from native fibers by an acid hydrolysis, giving rise to highly crystalline and rigid nanoparticles (often referred to as CNC or nanowhiskers) which are shorter (100s to 1000 nanometers) than the nanofibrils obtained through the homogenization, microfluiodization or grinding routes. The resulting material is known as nanocrystalline cellulose (NCC).

History and Terminology

The terminology microfibrillated/nanocellulose or (MFC) was first used by Turbak, Snyder and Sandberg in the late 1970s at the ITT Rayonier labs in Whippany, New Jersey, USA to describe a product prepared as a gel type material by passing wood pulp through a Gaulin type milk homogenizer at high temperatures and high pressures followed by ejection impact against a hard surface. The terminology (MFC) first appeared publicly in the early 1980s when a number of patents and publications were issued to ITT Rayonier on this totally new nanocellulose composition of matter. In later work Herrick at Rayonier also published work on making a dry powder form of the gel. Since Rayonier is one of the world's premier producers of purified pulps their business interests have always been 1) to create new uses and new markets for pulps and 2) never to compete with new or potentially new customers. Thus, as the patents issued, Rayonier gave free license to whoever wanted to pursue this new use for cellulose. Rayonier,as a company, never pursued scale-up. Rather, Turbak et al. pursued 1) finding new uses for the MFC/nanocellulose. These included using MFC as a thickener and binder in foods, cosmetics, paper formation, textiles, nonwovens, etc. and 2) evaluate swelling and other techniques for lowering the energy requirements for MFC/ Nanocellulose production. ITT closed the Rayonier Whippany Labs in 1983–84 and further work on making a dry powder form of MFC was done by Herric at the Rayonier labs in Shelton, Washington, USA

The field was later taken up in Japan in the mid 1990s by the group of Taniguchi and co-workers and later by Yano and co-workers. and a host of major companies (see numerous U.S. patents issued to P&G, J&J, 3M, McNeil, etc. using U.S. patent search under inventor name Turbak search base). Today, there are still extensive research and development efforts around the world in this field.

As of August 2012, NCC was produced in a factory operated by CelluForce, producing one ton per day, and by a just-opened facility operated by the US Forest Service. Recent developments (as of January 2016) show a renewed effort to industrialise MFC and find new applications fields with newly founded companies such as Borregaard's Exilva, FiberLean (a joint-venture of Imerys and Omya), American Process and WEIDMANN Fiber Technology among others.

Manufacture

Nanocellulose/CNF or NCC can be prepared from any cellulose source material, but woodpulp is normally used.

The nanocellulose fibrils may be isolated from the wood-based fibers using mechanical methods which expose the pulp to high shear forces, ripping the larger wood-fibres apart into nanofibers. For this purpose high-pressure homogenizers, ultrasonic homogenizers, grinders or microfluidizers can be used. The homogenizers are used to delaminate the cell walls of the fibers and liberate the nanosized fibrils. This process is responsible for the high energy consumptions associated with the fiber delamination. Values over 30 MWh/tonne are not uncommon.

Pre-treatments are sometimes used to address this problem. Examples of such pre-treatments are enzymatic/mechanical pre-treatment and introduction of charged groups e.g. through carboxymethylation or TEMPO-mediated oxidation. It has been shown that energy consumption can be heavily decreased using these pre-treatments and Lindström and Ankerfors has reported values below 1 MWh/tonne.

Cellulose nanowhiskers, a more crystalline form of nanocellulose, are formed by the acid hydrolysis of native cellulose fibers commonly using sulfuric or hydrochloric acid. The amorphous sections of native cellulose are hydrolysed and after careful timing, the crystalline sections can be retrieved from the acid solution by centrifugation and washing. Cellulose nanowhiskers are rodlike highly crystalline particles (relative crystallinity index above 75%) with a rectangular cross section. Their dimensions depend on the native cellulose source material, and hydrolysis time and temperature.

Recent breakthroughs in nanocellulose production were announced April 2013 by R. Malcolm Brown Jr. PHD, a biology professor at the University of Texas at Austin, who presented his team's findings at an American Chemical Society conference in New Orleans.

Borregaard AS, a Norwegian biorefinery company based in Sarpsborg, Norway, announced in October 2014 that they are investing in a facility to produce microfibrillated cellulose on a commercial scale. The production facility will have a capacity of 1000 tonnes per year, and is scheduled to be operational from Q3, 2016.

American Process Inc, a biorefinery technology development company headquartered in Atlanta, Georgia, has also recently announced a manufacturing process using sulfur dioxide and ethanol pretreatment of biomass to produce both cellulose nanofibrils and cellulose nanocrystals and lignin-coated, hydrophobic varieties of each.

ICAR-CIRCOT Nanocellulose Pilot Plant, a unique facility for production of nanocellulose from cotton and other agro-biomass, was inaugurated by Padma Visbhushan Dr. R. A. Mashelkar in Mumbai, India on August 21, 2015. This pilot plant was funded by World Bank supported National Agricultural Innovation Project of Indian Council of Agricultural Research, New Delhi. This pilot plant has the capacity of 10 kg nanocellulose production per shift of 8 hour operation and uses patented energy efficient and eco-friendly technologies. Both nanocrystalline cellulose and nanofibrillated cellulose can be produced in this plant (both in aqueous and dry form). This is established to act as an incubation center for new entrepreneurs and as technology demonstration unit.

Structure and Properties

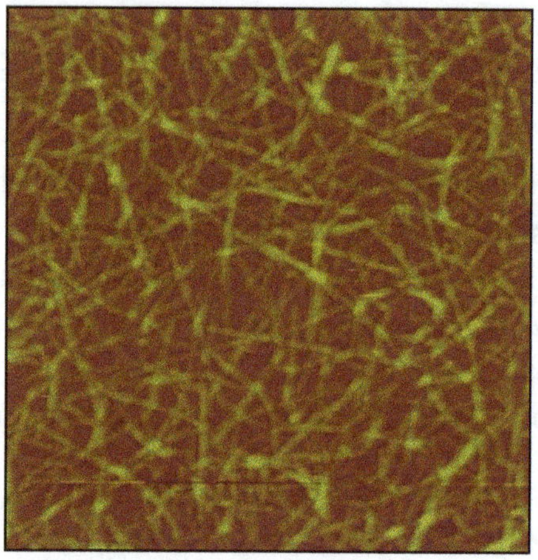

AFM height image of carboxymethylated nanocellulose adsorbed on a silica surface. The scanned surface area is 1μm2.

Nanocellulose Dimensions and Crystallinity

The ultrastructure of cellulose derived from various sources has been extensively studied. Techniques such as transmission electron microscopy (TEM), scanning electron microscopy (SEM), atomic force microscopy (AFM), wide angle X-ray scattering (WAXS), small incidence angle X-ray diffraction and solid state ^{13}C cross-polarization magic angle spinning (CP/MAS) nuclear magnetic resonance (NMR) spectroscopy have been used to characterize nanocellulose morphology. These methods have typically been applied for the investigation of dried nanocellulose morphology.

Although a combination of microscopic techniques with image analysis can provide information on nanocellulose fibril widths, it is more difficult to determine nanocellulose fibril lengths because of entanglements and difficulties in identifying both ends of individual nanofibrils. It has been reported that nanocellulose suspensions may not be homogeneous and that they consist of various structural components, including cellulose nanofibrils and nanofibril bundles.

Most methods have typically been applied to investigation of dried nanocellulose dimensions, although a study was conducted where the size and size-distribution of enzymatically pre-treated nanocellulose fibrils in a suspension was studied using cryo-TEM. The fibrils were found to be rather mono-dispersed mostly with a diameter of ca. 5 nm although occasionally thicker fibril bundles were present. It should be noted that, some newly published results indicated that by combining ultrasonication with an "oxidation pretreatment", cellulose microfibrils with a lateral dimension that belows 1 nm is observed by AFM. The lower end of the thickness dimension is around 0.4 nm, which is related to the thickness of a cellulose monolayer sheet.

The aggregate widths can be determined by CP/MAS NMR developed by Innventia AB, Sweden, which also has been demonstrated to work for nanocellulose (enzymatic pre-treatment). An average width of 17 nm has been measured with the NMR-method, which corresponds well with SEM and TEM. Using TEM, values of 15 nm have been reported for nanocellulose from carboxymethylated pulp. However, also thinner fibrils can also be detected. Wågberg et al. reported fibril widths of 5–15 nm for a nanocellulose with a charge density of about 0.5 meq./g. The group of Isogai reported fibril widths of 3–5 nm for TEMPO-oxidized cellulose having a charge density of 1.5 meq./g.

The influence of cellulose pulp chemistry on the nanocellulose microstructure has been investigated using AFM to compare the microstructure of two types of nanocellulose prepared at Innventia AB (enzymatically pre-treated nanocellulose and carboxymethylated nanocellulose). Due to the chemistry involved in producing carboxymethylated nanocellulose, it differs significantly from the enzymatically pre-treated one. The number of charged groups on the fibril surfaces are very different. The carboxymethylation pre-treatment makes the fibrils highly charged and, hence, easier to liberate, which results in smaller and more uniform fibril widths (5–15 nm) compared to the enzymatically pre-treated nanocellulose, where the fibril widths were 10–30 nm. The degree of crystallinity and the cellulose crystal structure of nanocellulose were also studied at the same time. The results clearly showed the nanocellulose exhibited cellulose crystal I organization and that the degree of crystallinity was unchanged by the preparation of the nanocellulose. Typical values for the degree of crystallinity were around 63%.

Viscosity

The unique rheology of nanocellulose dispersions was recognized by the early investigators. The high viscosity at low nanocellulose concentrations makes nanocellulose very interesting as a non-caloric stabilizer and gellant in food applications, the major field explored by the early investigators.

The dynamic rheological properties were investigated in great detail and revealed that the storage and loss modulus were independent of the angular frequency at all nanocellulose concentrations between 0.125% to 5.9%. The storage modulus values are particularly high (104 Pa at 3% concentration) compared to results for cellulose nanowhiskers (102 Pa at 3% concentration). There is also a particular strong concentration dependence as the storage modulus increases 5 orders of magnitude if the concentration is increased from 0.125% to 5.9%.

Nanocellulose gels are also highly shear thinning (the viscosity is lost upon introduction of the shear forces). The shear-thinning behaviour is particularly useful in a range of different coating applications.

Mechanical Properties

Crystalline cellulose has interesting mechanical properties for use in material applications. Its tensile strength is about 500MPa, similar to that of aluminium. Its stiffness is about 140–220 GPa, comparable with that of Kevlar and better than that of glass fiber, both of which are used commercially to reinforce plastics. Films made from nanocellulose have high strength (over 200 MPa), high stiffness (around 20 GPa) and high strain (12%). Its strength/weight ratio is 8 times that of stainless steel.

Barrier Properties

In semi-crystalline polymers, the crystalline regions are considered to be gas impermeable. Due to relatively high crystallinity, in combination with the ability of the nanofibers to form a dense network held together by strong inter-fibrillar bonds (high cohesive energy density), it has been suggested that nanocellulose might act as a barrier material. Although the number of reported oxygen permeability values are limited, reports attribute high oxygen barrier properties to nanocellulose films. One study reported an oxygen permeability of 0.0006 (cm^3 µm)/(m^2 day kPa) for a ca. 5 µm thin nanocellulose film at 23 °C and 0% RH. In a related study, a more than 700-fold decrease in oxygen permeability of a polylactide (PLA) film when a nanocellulose layer was added to the PLA surface was reported.

The influence of nanocellulose film density and porosity on film oxygen permeability has recently been explored. Some authors have reported significant porosity in nanocellulose films, which seems to be in contradiction with high oxygen barrier properties, whereas Aulin et al. measured a nanocellulose film density close to density of crystalline cellulose (cellulose Iß crystal structure, 1.63 g/cm^3) indicating a very dense film with a porosity close to zero.

Changing the surface functionality of the cellulose nanoparticle can also affect the permeability of nanocellulose films. Films constituted of negatively charged cellulose nanowhiskers could effectively reduce permeation of negatively charged ions, while leaving neutral ions virtually unaffected. Positively charged ions were found to accumulate in the membrane.

Multi-Parametric Surface Plasmon Resonance is one of the methods to study barrier properties of natural, modified or coated nanocellulose. The different antifouling, moisture, solvent, antimicrobial barrier formulation quality can be measured on the nanoscale. The adsorption kinetics as well as the degree of swelling can be measured in real-time and label-free.

Foams

Nanocellulose can also be used to make aerogels/foams, either homogeneously or in composite formulations. Nanocellulose-based foams are being studied for packaging applications in order to replace polystyrene-based foams. Svagan et al. showed that nanocellulose has the ability to reinforce starch foams by using a freeze-drying technique. The advantage of using nanocellulose instead of wood-based pulp fibers is that the nanofibrills can reinforce the thin cells in the starch foam. Moreover, it is possible to prepare pure nanocellulose aerogels applying various freeze-drying and super critical CO2 drying techniques. Aerogels and foams can be used as porous templates. Tough ultra-high porosity foams prepared from cellulose I nanofibrill suspensions were studied by Sehaquiet al. A wide range of mechanical properties including compression was obtained by controlling density and nanofibrill interaction in the foams. Cellulose nanowhiskers could also be made to gel in water under low power sonication giving rise to aerogels with the highest reported surface area (>600m2/g) and lowest shrinkage during drying (6.5%) of cellulose aerogels. In another study by Aulin et al., the formation of structured porous aerogels of nanocellulose by freeze-drying was demonstrated. The density and surface texture of the aerogels was tuned by selecting the concentration of the nanocellulose dispersions before freeze-drying. Chemical vapour deposition of a fluorinated silane was used to uniformly coat the aerogel to tune their wetting properties towards non-polar liquids/oils. The authors demonstrated that it is possible to switch the wettability behaviour of the cellulose surfaces between super-wetting and super-repellent, using different scales of roughness and porosity created by the freeze-drying technique and change of concentration of the nanocellulose dispersion. Structured porous cellulose foams can however also be obtained by utilizing the freeze-drying technique on cellulose generated by Gluconobacter strains of bacteria, which bio-synthesize open porous networks of cellulose fibers with relatively large amounts of nanofibrills dispersed inside. Olsson et al. demonstrated that these networks can be further impregnated with metalhydroxide/oxide precursors, which can readily be transformed into grafted magnetic nanoparticles along the cellulose nanofibers. The magnetic cellulose foam may allow for a number of novel applications of nanocellulose and the first remotely actuated magnetic super sponges absorbing 1 gram of water within a 60 mg cellulose aerogel foam were reported. Notably, these highly porous foams (>98% air) can be compressed into strong magnetic nanopapers, which may find use as functional membranes in various applications.

Surface Modification

The surface modification of nanocellulose is currently receiving a large amount of attention. Nanocellulose displays a high concentration of hydroxyl groups at the surface which can be reacted. However, hydrogen bonding strongly affects the reactivity of the surface hydroxyl groups. In addition, impurities at the surface of nanocellulose such as glucosidic and lignin fragments need to be removed before surface modification to obtain acceptable reproducibility between different batches.

Cellulose nanofibers can also be modified as cationic. This cationic cellulose increase the affinity for anions

Safety Aspects

Health, safety and environmental aspects of nanocellulose have been recently evaluated. Processing of nanocellulose does not cause significant exposure to fine particles during friction grinding or spray drying. No evidence of inflammatory effects or cytotoxicity on mouse or human macrophages can be observed after exposure to nanocellulose. The results of toxicity studies suggest that nanocellulose is not cytotoxic and does not cause any effects on inflammatory system in macrophages. In addition, nanocellulose is not acutely toxic to Vibrio fischeri in environmentally relevant concentrations.

Applications

The properties of nanocellulose (e.g. mechanical properties, film-forming properties, viscosity etc.) makes it an interesting material for many applications and the potential for a multibillion-dollar industry.

Paper and Paperboard

There is potential of nanocellulose applications in the area of paper and paperboard manufacture. Nanocelluloses are expected to enhance the fiber-fiber bond strength and, hence, have a strong reinforcement effect on paper materials. Nanocellulose may be useful as a barrier in grease-proof type of papers and as a wet-end additive to enhance retention, dry and wet strength in commodity type of paper and board products.

Composite

As described above the properties of the nanocellulose makes an interesting material for reinforcing plastics. Nanocellulose has been reported to improve the mechanical properties of, for example, thermosetting resins, starch-based matrixes, soy protein, rubber latex, poly(lactide). The composite applications may be for use as coatings and films, paints, foams, packaging.

Food

Nanocellulose can be used as a low calorie replacement for today's carbohydrate additives used as thickeners, flavour carriers and suspension stabilizers in a wide variety of food products and is useful for producing fillings, crushes, chips, wafers, soups, gravies, puddings etc. The food applications were early recognised as a highly interesting application field for nanocellulose due to the rheological behaviour of the nanocellulose gel.

Hygiene and Absorbent Products

Different applications in this field include:

- Super water absorbent (e.g. for incontinence pads material)
- Nanocellulose used together with super absorbent polymers
- Use of nanocellulose in tissue, non-woven products or absorbent structures
- Use as antimicrobial films

Emulsion and Dispersion

Apart from the numerous applications in the area of food additives, the general area of emulsion and dispersion applications in other fields has also received some attention. Oil in water applications were early recognized. The area of non-settling suspensions for pumping sand, coal as well as paints and drilling muds was also explored by the early investigators.

Oil Recovery

Hydrocarbon fracturing of oil-bearing formations is a potentially interesting and large-scale application. Nanocellulose has been suggested for use in oil recovery applications as a fracturing fluid. Drilling muds based on nanocellulose have also been suggested.

Medical, Cosmetic and Pharmaceutical

The use of nanocellulose in cosmetics and pharmaceuticals was also early recognized. A wide range of high-end applications have been suggested:

- Freeze-dried nanocellulose aerogels used in sanitary napkins, tampons, diapers or as wound dressing
- The use of nanocellulose as a composite coating agent in cosmetics e.g. for hair, eyelashes, eyebrows or nails
- A dry solid nanocellulose composition in the form of tablets for treating intestinal orders
- Nanocellulose films for screening of biological compounds and nucleic acids encoding a biological compound
- Filter medium partly based on nanocellulose for leukocyte free blood transfusion
- A buccodental formulation, comprising nanocellulose and a polyhydroxylated organic compound
- Powdered nanocellulose has also been suggested as an excipient in pharmaceutical compositions
- nanocellulose in compositions of a photoreactive noxious substance purging agent
- Elastic cryo-structured gels for potential biomedical and biotechnological application.
- Matrix for 3D cell culture

Other Applications

- Activate the dissolution of cellulose in different solvents
- Regenerated cellulose products, such as fibers films, cellulose derivatives
- Tobacco filter additive
- Organometallic modified nanocellulose in battery separators
- Reinforcement of conductive materials
- Loud-speaker membranes
- High-flux membranes
- Flexible electronic displays
- Computer components
- Capacitors
- Lightweight body armour and ballistic glass
- Corrosion inhibitors

References

- Pacios Pujadó, Mercè (2012). Carbon Nanotubes as Platforms for Biosensors with Electrochemical and Electronic Transduction. Springer Heidelberg. pp. XX,208. doi:10.1007/978-3-642-31421-6. ISBN 978-3-642-31421-6. ISSN 2190-5053.

- Lindström, Tom; Mikael Ankerfors (2009). "NanoCellulose Developments in Scandinavia". 7th International Paper and Coating Chemistry Symposium (Preprint CD ed.). Hamilton, Ontario: McMaster University Engineering. ISBN 978-0-9812879-0-4.
- Chinga-Carrasco, G., Miettinen, A., Luengo Hendriks, C.-L., Gamstedt, E.K., Kataja, M. (2011). Structural Characterisation of Kraft Pulp Fibres and Their Nanofibrillated Materials for Biodegradable Composite Applications. InTech. ISBN 978-953-307-352-1.
- "Green sonochemical synthesis of silver nanoparticles at varying concentrations of κ-carrageenan" (PDF). springer.com. Retrieved 15, February 2020.
- Kolosnjaj J, Szwarc H, Moussa F (2007). "Toxicity studies of carbon nanotubes". Adv Exp Med Biol. Advances in Experimental Medicine and Biology. 620: 181–204. doi:10.1007/978-0-387-76713-0_14. ISBN 978-0-387-76712-3. PMID 18217344.
- Godovsky, D. Y. (2000). "Device Applications of Polymer-Nanocomposites". In Chang, J. Y. Biopolymers · PVA Hydrogels, Anionic Polymerisation Nanocomposites. Advances in Polymer Science. 153. pp. 163–205. doi:10.1007/3-540-46414-X_4. ISBN 978-3-540-67313-2.
- Berglund, Lars (2005). "Cellulose-based nanocomposites". In A.K. Mohanty; M. Misra; L. Drzal (Eds). Natural fibers, biopolymers and biocomposites. Boca Raton, Florida: CRC Press. pp. 807–832. ISBN 978-0-8493-1741-5.

5

Nanometrology: An Integrated Study

Nanometrology focuses on the science of measurement at the nanoscale level. It plays an important role in the production of nanomaterials and devices accurately within the field of nanomanufacturing. This chapter has been carefully written to provide an easy understanding of the varied facets of nanometrology as well as nanometrology measurement techniques.

Nanometrology involves the measurement of geometrical features of size, shape and roughness at the nanoscale. Though not an innate aspect of the specimen under study, these geometrical features are often measured against an arbitrarily fixed co-ordinate system, more so in an engineering application. With advent of new age metrological instruments (e.g. SPM), the parameters of measurement also include a physical quantity (e.g. force). Another important aspect that has to be borne in mind when the scale of objects reduces is the relative importance or relevance of the physical features namely size, shape and roughness. Below are three plots that illustrate the relative importance of the physical features as the scale reduces.

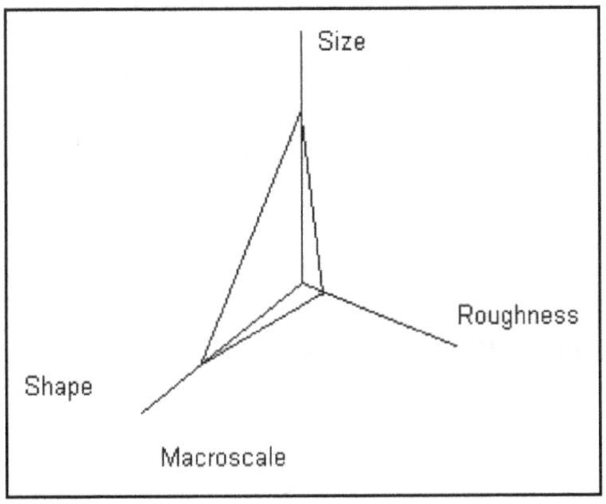

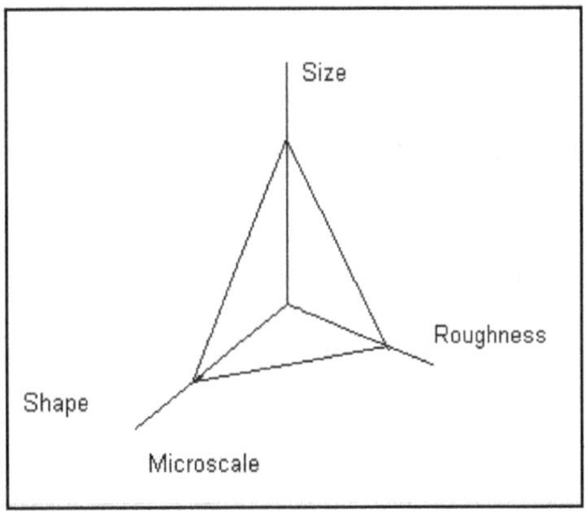

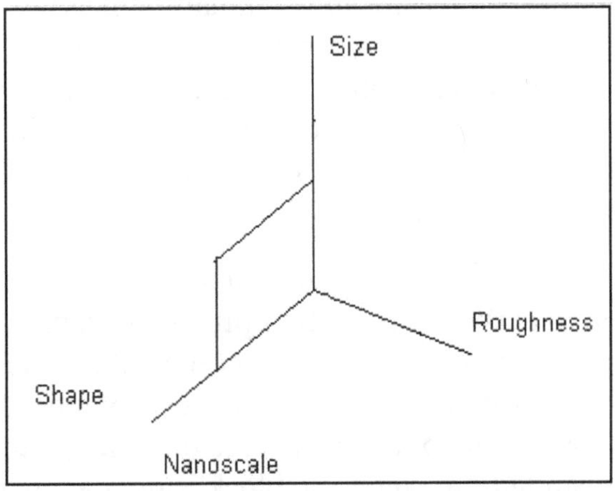

Figure: Relative relationship between different geometric components in conventional, micromechanics & quantum engineering.

Macro scale measurements are usually encountered in conventional engineering applications; micro scale measurements are encountered in micro mechanics while nanoscale measurements are encountered in quantum engineering. Also, as scale reduces additional physical constraints such as tunneling and other such related phenomenon are also to be considered. The above plots, in a crude way, explain how the very definition of roughness changes as we reduce the scale.

At the Nanoscale

In conventional, macro scale engineering, the size, shape and roughness are purely functions of manufacture process mechanisms like time of machining, depth of cut and the path, which the cutting tool adopts. Loss of co-ordination between any of these parameters leads to large errors. But in nanoscale or at molecular level, the scenario is rather strange. The size and shape are themselves so small that, roughness as a separate physical quantity is not definable.

For the sake of demarcating various engineering materials, the whole spectrum has been split into four levels.

a) Bulk (μm): This covers all systems right from micro electro mechanical systems (MEMS) to large structures.

b) Particles (100nm): It includes powders and composites, for example aluminium.

c) Micro clusters (10nm): It includes alloys, catalysts, nanotubes, and fullerenes.

d) Molecular scale (1nm):It encompasses thin films such as Langmuir-Blogett and self- assembly proteins.

Geometric features and the scale of size:

The total geometry of an object can be mathematically represented as,

$$G = S_i \cap S_{sh} \cap R$$

The sanctity in representing the geometry in such a manner is that it represents the nominal independence of the physical features. For e.g. in conventional engineering, the size (Si) is much greater than shape (S_{sh}) and roughness (R) and is hence nominally independent of it. In other words, the values of shape and roughness do not alter the size value. In the same way, we can observe that as the size factor decreases there is growing influence of shape and roughness. As the scale reduces, it becomes progressively difficult to describe size, shape and roughness independently. This apparent arises due to difficulty in scaling down roughness.

Reduction in size is pretty simple for there is only a general reduction in all dimensions and there is not a large variation to make the reduction process so tedious. In a similar fashion, the shape is more dependent on, as already mentioned on the path, which the cutting tool takes up and is dependent on parameters that need not be rigorously balanced. To put it more perspicuously, while it does not require much approximation to measure a rectangular block, but it does requires assumption of a mean length, when the block is only a few microns in length and has rather rugged ends.

Force Balance Variation with Scale

As the scale of size is reduced, there is profound change in the relative importance of the components of the force equations. At the mm level and above, the inertial component dominates. However, as the scale is reduced to the nanometre level, the inertial component falls off faster in magnitude than either the damping or elastic terms. The most important component becomes the elastic force. However, elastic properties are controllable since the material properties are known. The inertial term is negligible leaving the damping term which is areal and is consequently dependent on the surface finish. Adhesion and friction depend a great deal on the relatively uncontrolled surface. The change in the relative importance of the forces is the reason why insects have such different movement capabilities than mammals. Insects can change direction instantly indicating little inertia. They can walk upside down on ceilings indicating that adhesion forces are much greater than force on mass.

Dependence of Roughness on Scale

Roughness at nanoscale can occur in three broad cases,

1) On large objects such as mirrors
2) On small objects such as in micro dynamics
3) At the molecular and atomic scale

On large objects, roughness has been regarded as irrelevant and the only probable effect it can have is on the reactivity at the surface due to increased surface area. This might probably the only limit that can be set for roughness in macro scale.

At cluster level, the priorities change. The clusters themselves are made up of few hundreds of atoms and more so, the whole functionality of the cluster lies at the surface than at the bulk. Thus what we might call roughness in macro level actually turns out to be shape at cluster level. To resolve this anomaly, the definition of roughness has been changed. While shape is defined as the configuration of molecules (that is automatically generated), the roughness is defined as the defects that arise during shape generation.

Turning our attention to the molecular level, roughness usually means the disorder in the layers of whiskers of carbon nanotubes and fibre, which could easily be avoided by careful processing and preparation techniques. Except in cases where the roughness is present at the substrate interface and the effect it causes is very much similar to what roughness might cause at macro level. Another common observation though not strictly classified, as roughness is the forming of oxides, sulphides that depends on the electron structure at the surface and hence comes under the classification of shape rather than roughness. Other such related observations are that of formation of a separate symmetry at surface of semiconductors, which produces a new unit cell at the surface called as surface reconstruction' and is a consequence of the structure seeking low potential energy. This surface reconstruction affects deposition of metal on the surface. Other phenomena include formation of surface colloids, coverage of surface with biological molecules like proteins and enzymes to impart activity etc. All these interesting applications just go on to show that roughness which is more of than not pejorative at macro scales turns out to be rather useful at nanoscale.

Surface and Bulk Properties in Nanoscale:

Roughness in the conventional sense as we know makes little sense at nano levels. More specifically, priority should be given to how well controlled is the production process to attain the desired reactance and stability rather than on attaining smooth polished surface. The surface starts playing so important roles that, cleavage have to be done in vacuum environments to maintain reactance and stability. Exposure to atmosphere can result in adsorption of chemicals onto the surface, diffusion of these chemicals into the bulk, alteration of electric and magnetic properties and so on. This is plainly the reason why roughness was defined as the defects present on the surface.

Defects rather than roughness take over as the important surface characteristics in the nanometer and sub-nanometer domains.

Defects are usually characterized by their physical dimensions like zero-dimensional defects (point defects), one-dimensional defects (dislocations), two-dimensional defects (slip and twin) and three-dimensional defects (voids and cracks). However, the maximum dimension of any defect is purely dependent on that of the parent sample. What we call a 3-D defect at nanoscale becomes a point defect at macro scale and hence the break in definition occurs at the nanoscale.

Engineering Shape

There are a number of shapes in engineering which are especially relevant. Of these, roundness is probably the most important. Practically every dynamic application involves rotation. The measurement and control of roundness is well understood at the macro level. However, as the size reduces it becomes more difficult, at the millimetre level and below, to include MEMS. The measurement is tricky. This is mainly due to centering. In addition, the signal as seen by the instrument is not the same as that obtained at the macro level.

For machined objects, shape is usually determined by the movement of the tool. This involves movement in two or three dimensions. The curve produced in space is usually simple, conic forms such as parabolas and spheres are often required.

Molecular Shape and Scale of Size

The aspect of shape at the nanoscale is that it can have a meaning i.e. it is predetermined by energy or other considerations. Whereas in engineering the shape usually determines whether the work-piece can move (i.e. roundness can determine how well a shaft revolves in a bearing). At the nanoscale, the shape of an object can be a result of the internal energy balance between, say, chemical and elastic energies. In clusters or even liquid droplets, the energy balance is between Coulomb attraction and surface energy. From these considerations, it is clear that in nanotechnology shape is often an output from energy considerations rather than an output to an engineering function. Shape can therefore have completely different implications in macro- as compared with nano- regimes.

In molecular domain, shape is determined by the chemical composition and the structures have a fixed shape depending on the particular molecular configuration. In other words, the shape of a structure is preordained. The question is whether it is inhibited by defects or not. Shapes are discrete, not continuous as in macro engineering. The measurement of shape therefore is usually not the issue: failure to achieve the expected shape is.

In quality control terms in the nano- and molecular regimes shape is often an attribute and not a variable, e.g. is it the correct shape rather than the 'value' of the shape. The shape of interfaces can also influence the quantum behavior of materials. The solution of Schrodinger's electron wave equation can be influenced by the boundary conditions; an ideal parabolic shape for the potential barrier in a quantum well depends on the shape of the boundary, which could be an interface or a surface.

Carbon Molecular Shapes

Due to their extremely precise shapes and sizes, carbon nanotubes and similar structures such as fullerenes have such remarkable electronic and mechanical properties. In principle, they can be used in gears, rack and pinion. These structures relate to the parent planar graphite.

Some enhanced mechanical properties include better stability, strength, stiffness and elastic modulus. The helicity, which is the arrangement of the carbon hexagons on their surface layer honeycomb lattice, is determined by the symmetry of the structure, the helix angle and the diameter of the tube (i.e. the shape and size).

It should be noted that it is not only carbon which has the potential for forming tubes, but also Molybdenum disulphide (MoS_2) the well-known lubricant also has a similar structure.

Thus to sum it up - metrological factors change in some way or the other as the size reduces below the nanometre level. The balance of forces also changes with size, with each component changing disproportionately. However, the metrological components change form rather than value. For the purely mechanical applications in microdyanmics, roughness, shape (form) and size become scrambled. Forces are small enough so that we need not consider the sub-surface stress and waviness.

Figure above shows the diversity of the geometric relationships. However there is inherent difficulty in establishing a common framework of standards to back the different types of measurements. Also, as the size decreases the nature of the signal changes. What appears as a straightforward deterministic measurement in engineering gets replaced by spatial statistical averaging and, finally, temporal statistics brought about by quantum mechanics.

Metrology at Nanoscale

Even after monitoring carefully how various geometrical features and force balance vary in their relative meaning significance, we face yet another problem and that arises from the instrument we use.

Stability of Signals

Length

The system shown is conventional engineering and is stable to the signal S. Uncertainty is so small that noise is negligible. The signal is acceptable. In (b) the actual signal at any point changes with position so, to get a meaningful signal S, the geometry has to be smoothed by integration giving m1 and m2. The signal is now S = m1- m2 where m1 and m2 are mean planes. The distance S can be accurate to much less than a nanometre because of the spatial integration.

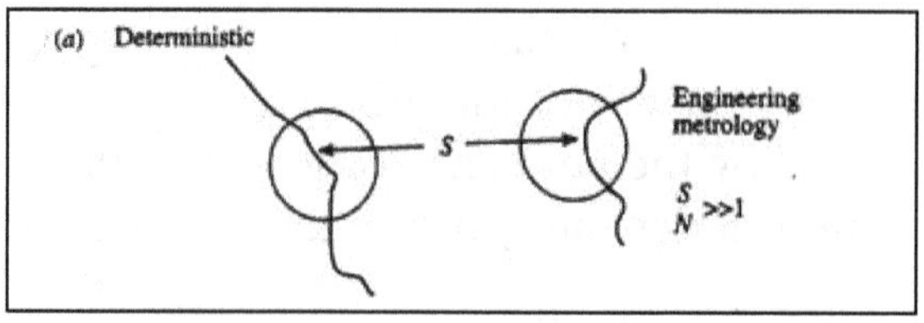

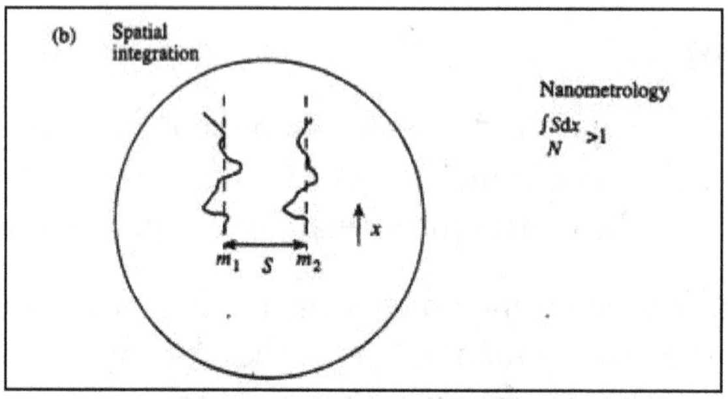

Figure: Signal form in (a) Engineering Metrology (b) Nanometrology.

In (c) at molecular and atomic levels the available signals passing between the two points is limited by the distance between points p1 and p2 which are molecular or atomic un size. In this regime tunneling across the gap or through a barrier is subject to quantum laws. The only way to increase the probability of an electron or photon moving between p1 and p2 and so closing the measurement loop is to increase the observation time (i.e. to integrate temporally). This automatically makes the instrument sensitive to the environment. This temporal problem is not the same as the bandwidth of electrical noise.

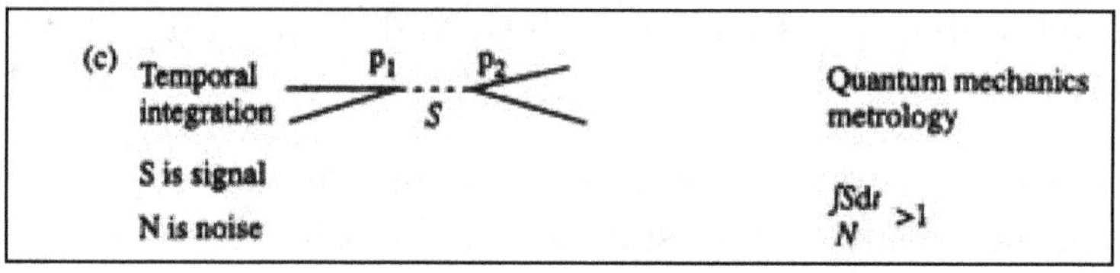

Figure: Quantum Mechanics Metrology.

The increase in resolution and accuracy of modern instruments forced by application pressure has brought their specification into the nano regime-in effect almost the same scale of size spatially as that of the surface roughness. This is why the subjects of surface metrology and nanometrology are blending and why position and spacing are being lumped into the new discipline of surface nanometrology.

Nano Position Sensing

This is defined as the technology of moving and measuring with sub-nanometre precision and has values, which are comparable to very fine surface finish. It is also obviously related to length. Optical methods can be used to measure position using white light fringes. However, a very common method is the use of piezoelectric crystals suitably energised by a known voltage to produce the movement, and some gauge, usually capacitative to control the movement. The control is necessary because, although piezoelectric crystals are very stiff, hey suffer greatly from non-linearity and hysteresis. It needs a very precise and accurate gauge to find out where the end of the crystal is in space.

As there are three components involved in the geometry of a work piece each having its idiosyncrasy it is wise to review the calibration procedure in order to see if three can be brought together in a traceable way at the nanoscale.

Calibration

General

The real problem lies in the way engineering surface metrology and microscopy have evolved. In many engineering applications such as those which occur in tribology have more emphasis on the height variations than on spacing or lateral dimensions. They are mainly observed for aerial structure and position in the plane of the object.

There are three types of length calibration artifacts:

1) Line spacing: In this type of calibration the effect of the shape of the probe and its interaction with the surface are small because any aberration introduced due to the interaction is common to both lines if measured from the same direction.
2) Line width.
3) Size or extension.

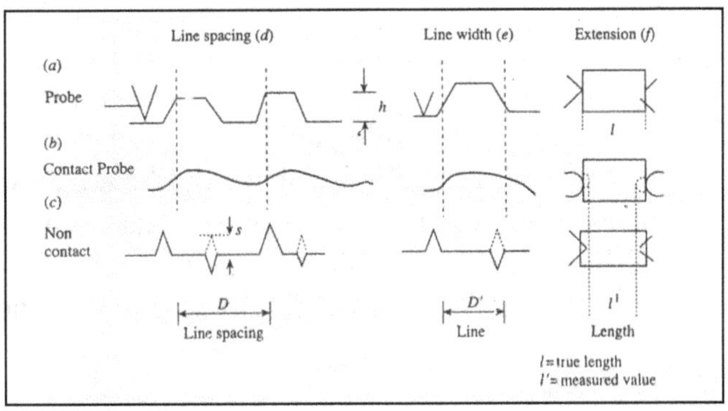

Figure: Length standards showing the effect of interaction between the probe and the artifact.

Probe

The probe is the most important component of the new generation scanning microscopes. The process of manufacturing a tip of a particular dimension is difficult. The most common example that can be given in this case is electrochemical etching techniques to fabricate tungsten and platinum/ iridium tips for use in STM. These tips have been made with a very high aspect ratio. The main idea of developing such a tip is to obtain a very small angle at the surface. In addition to this by this process it is also possible to keep the radius of curvature very small. The technique of electrochemical etching to make STM is a natural progression from field ion microscopy. We would prefer to use platinum for tips rather than the common tungsten is due to the fact that platinum, although it is softer material, platinum is inert to oxidation. On the other hand tungsten would easily get oxidized to WO2. We would also introduce some amount of iridium into the tip surface as it would tend to make the tip much harder and stiffer so that it can withstand much higher temperatures without undergoing any amount of plastic deformation.

Some of the Unconventional Methods of Stm Tip Preparation Ion Sputter Thinning Method

The ion sputter thinning method is used to get tips which would offer a better resolution and reliability. These tips are capable of achieving atomic resolution spatially can be fabricated. In this process the given material should be made into the dimensions in the micron level. This starting material can be obtained using the ultra-microtome cutting operation. However we should choose the appropriate material for cutting or shearing of the initial sample. Generally diamond coated tungsten carbide blades are used for this purpose. This cut sample which is obtained is then sputtered using an ion bean of Ar (Ar sputtering process). The main disadvantage of this process is that it is very slow and it would take about 18 hrs.

The novel and simple method is to develop a tip fabricated out of pencil lead by coating tungsten tips with colloidal graphite. There is also another method of tip fabrication of tips which involves the shearing of Tt/ Ir wire. By shearing the wire at a very large angle the wire gets a cut tip which is often good enough to use in measuring roughness in micro-areas.

Use of Carbon Nanotube

One of the most novel methods is to use carbon nanotubes as a probe for STM. These would be very suitable for nanometrology. This is because the carbon nanotubes have a very small diameter, a large aspect ratio (length/ diameter). In addition to this they have the property of high stiffness to lateral movement. This very important because when we apply very large potential through the tip for rastering patterns then the tip would tend to vibrate in the direction parallel to the surface movement. This would lead to misalignment of the probe. The carbon nanotubes have high tensile strength along the axial direction which would contribute to the stiffness. There is also scope for using this carbon nano tube as a nanopipette also along with STM probe usage. Through the hollow section of the tube it is possible to send few molecules or atoms to the particular region which we have just probed. The main difficulty is that of attaching them to the body of the main pick up part.

There are also optical super tips which can be smaller than 10nm. By the use of these optical super tips is possible that only single molecules that can absorb light energy and transcribe to different optical properties. This type of tip is used in Scanning Near-Field Optical Microscope (SNOM) which is a scanning probe microscope that allows optical imaging with spatial resolution beyond the diffraction limit. A nanoscopic light source, usually a fiber tip with an aperture smaller than 100 nm, is scanned much closed to the sample surface in the region called "optical near-field". Particularly, scanning in liquids becomes easily feasible, which is vital for biological applications where the sample is always in a solution medium. For example, its applications reach from routine control of micro contact-printed samples over the determination of the orientation of proteins on surfaces to studies on single fluorescent molecules. This technique makes the sub-100 nm length scale accessible to optical investigations with all the benefits of different contrast mechanisms such as fluorescence and polarization, including the option of chemical identification. Additionally it has the benefit of recording the sample topography simultaneously with the optical information alike an atomic force microscope.

Special silicon tips capable of being used in the fabrication of semiconductor devices. For example by atomic force microscopy which involves small size movements of atom by atom, resulting in possible locations of 1014 GB/m2 on silicon wafers. The silicon tips are generally highly acicular in nature having a tip angle of only about 8-100. They are based on silicon beams and quartz tips, which are completely coated with a thin aluminum film. Despite the fact that no physical opening is created at the tip's apex prior to scanning, the energy throughput is sufficiently high and an optical resolving power of around 32 nm has been shown.

Height Measurement Calibration at the Nanometer Scale of Size

One of the big problems associated with the height calibration at the nanometer level is that there is a tendency to try to relate the calibration procedure devised for contact methods with those of optical methods. In particular the optical methods have been used to measure the artifacts developed for stylus contact methods (note that this attempted comparison has been carried out for roughness standards. If the surface is rough the optical and mechanical probe gives the surface geometry fairly adequately. The difference between optical and mechanical techniques is not usually significant. But if the surface is smooth then even small differences can cause a serious divergence of the two methods.

Although it seems sensible to use optical interferometry to measure position, displacement and step height, typical results with interferometry where d is the measured height would give an uncertainty of measurement of 10^{-4}d nm + 2.5nm whereas the stylus method gives 1.5x10^{-3}d nm + 0.7nm. The errors associated with the stylus method or nanostep method can be attributed to small non-linearities ion the transducers, whereas the errors in the optical methods are usually attributed to difficulties in fringe position measurement, surface contamination and roughness of the surface of the steps.

The true way to compare instruments properly is to measure along exactly the same track by using kinematic methods together with track identification techniques for example the double cross method as shown in the figure.

Using the relocation method there is no ambiguity of trace position. The yaw angle is adjusted until just the two marks appear on the trace. There is sometimes the possibility of occurrence of noise. This occurrence of noise may lead to the burial of the signal because there is a mismatch between what is being measured and the metrology unit. One example of this mismatch is the use of light having nominal wavelength of 0.5μm to measure nano detail. The mismatch in this case is about 500:1. This gives a very low signal to noise ratio. Also there is a limit to which what can be extracted from the attenuated signal.

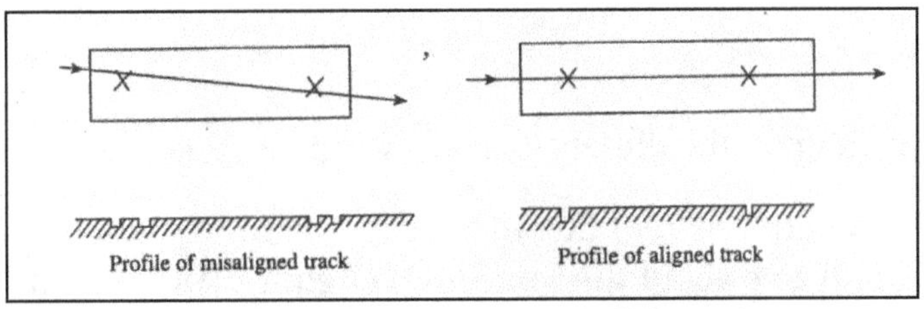

Figure: Relocation method for comparing different instruments.

X-ray Interferometer

In this arrangement the interferometer is a monolithic block of silicon. This is machined to project three blades from one face. The third blade is a part of the ligament hinge. X- rays enter from the left and are focused via the second blade B onto C which is the movable blade. The movement of C varies with the absorption of X- rays by the blade. This is then detected by a scintillation counter.

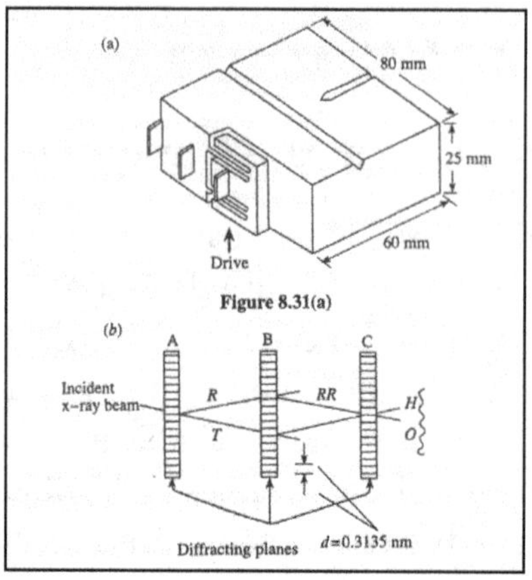

Figure: Interferometer structure (b) Arrangement of silicon blocks inside the setup.

If movable blade C is moved linearly upward the voltage output varies sinusoidally. This signal is the master which, when related back to the lattice spacing can be used to compare with the test transducer signal whose contact point or focal point is positioned on the top of the moving ligament hinge configuration of the crystal. The actual value of the x-ray wavelength is not critical but the most important factor is the positioning of the parallel faces of the blades11. Interpolation of the sinusoidal signal would give much better sub-nanometer accuracy and resolution.

Noise

The most common source of noise in many instruments is due to electronic or thermal effects.

Electronic

This is due to the electronic fluctuations inherent in the input stage of the amplifier or the probe itself. This noise is also commonly referred to as Johnson noise and Shot noise. The equation for RMS voltage for Johnson noise is,

$$E\left(V_j^2\right) = KT\Delta BR^{-1}$$

ΔB = bandwidth gap

K= constant which depend upon the dimensions of the tip

Shot noise is the variation in the current due to the random movement of electrons. It has a same bandwidth ΔB as above and give about the same level of equivalent noise as Johnson noise.

Thermal Effects

The deflection of the end of an AFM cantilever beam may be caused by the Brownian motion. For a simple detection system based on viewing the end of the cantilever, the expected variation in z is given by,

$$E\left(\Delta z^2\right) = 0.065\,K^{-0.5}$$

$$K = 0.25\,E\,wh^2\,L^{-3}$$

Where h is thickness, w is width and L is the length of the cantilever.

In cases where the deflection is measured by a laser beam reflected from the cantilever end, an optical cantilever, then dz(L)/dx is the operative measurand.

$$Z(L) = 0.667\,L\,dz(L)/dx$$

By applying thermal vibration theory to cover the two cases when (a) the one end of the cantilever is free and (b) when it is making contact with the surface it is possible to deduce the different modes of vibration of the beam using standard above given formulae.

When all the vibration modes are taken into account the root mean square deflection $\sqrt{z^2}$

is given that the optical lever is used for measurement,

$\sqrt{(4kt/rK)}$ For free end

$\sqrt{(kt/3rK)}$ For contact end

Where k is boltzman's constant and t is temperature,

For a temperature of 22 °C and a spring constant such as given by K in Nm the values of the resultant thermal noise are just less than 10-10m. This means that the mechanical noise and the electronic noise are about the same and are uncomfortably close to the desired resolution. For small structures ex small cantilevers for AFM, the internal friction has been broken down into surface effects and bulk effects, both of which dissipate energy independently.

Calibration Artefacts

It is always very difficult to provide specimens with which to calibrate precision instruments. In engineering applications the obvious example of the use of artifacts is the use of gauge blocks, which are used as a practical intermediary between the test piece and length standards whose values can be traced back to the standard meter. However these gauge block concept cannot be extended to ranges of micrometer and nanometer.

One of the problems is the fact that there is a need for an artifact of regular pattern. This type is needed for nano-structure spacing as well as for lateral information in scanning microscopes and for filter characteristics in micro-roughness. The most common technique for measuring the roughness of the artifacts is by diffraction grating. The diamond turning is used to generate sine waves in copper. These are usually chrome plated to give hard-wearing properties12. However this technique cannot be used as effectively used for stylus or optical methods. This is because the stylus method tends to ignore remnant turning marks in the surface and light in optical techniques gets diffracted.

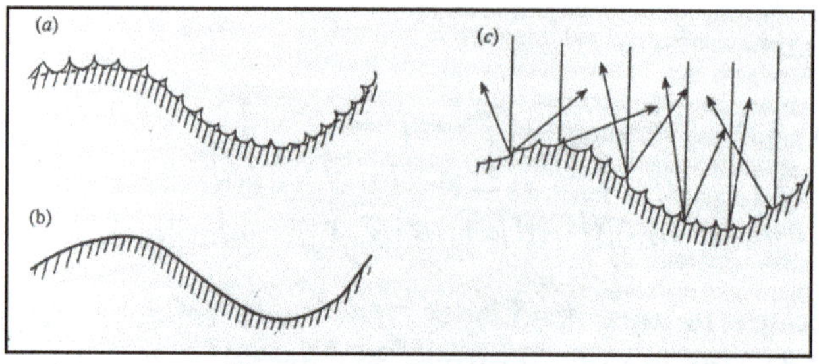

Figure: Attempt to calibrate optical instrument with tactile standard.

Nanometrology Measurement Techniques

Atomic Force Microscopy (AFM)

An atomic force microscope is a type of high resolution scanning probe microscope that has a resolution that you can measure in fractions of a nanometer.

It was pioneered in 1986 by Nobel Prize Winner Gerd Binnig along with Calvin Quate and Christoph Gerber.

One of the most important tools for imaging on the nanometer scale, Atomic Force Microscopy uses a cantilever with a sharp probe that scans the surface of the specimen.

When the tip of the probe travels near to a surface, the forces between the tip and sample deflect the cantilever according to Hooke's law.

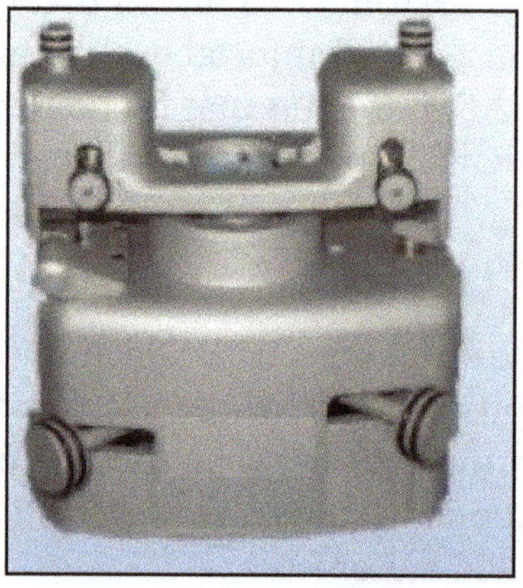

Atomic force microscopy will measure a number of different forces depending on the situation and the sample that you want to measure.

As well as the forces, other microscopes can include a probe that performs more specialized measurements, such as temperature.

The force deflects the cantilever, and this changes the reflection of a laser beam that shines on the top surface of the cantilever onto an array of photodiodes. The variation of the laser beam is a measure of the applied forces.

Contact and Non-contact Modes

There are two primary modes of operation for an atomic force microscope, namely contact mode and non-contact mode depending on whether the cantilever vibrates during the operation.

In contact mode, the cantilever drags across the sample surface and it uses the deflection of the cantilever to measure the contours of the surface.

To eliminate the noise and drift that can affect a static signal, low stiffness cantilevers are used, but this allows strong attractive forces to pull the tip to the surface. To eliminate this attraction, the tip is in contact with the surface where the overall force is repulsive.

In non-contact mode, the tip vibrates slightly above its resonance frequency and does not contact the surface of the sample. Any long range forces, like van der Waals forces, decreases the resonant frequency of the cantilever.

A feedback loop system helps to maintain the oscillation amplitude constant by changing the distance from the tip to the sample. Recording the distance between the tip and sample at each point allows the software to construct a topographic image of the sample surface.

Most samples will form a layer of moisture on the surface if stored at ambient conditions, and this can make it difficult to measure the sample accurately.

If the probe tip is close enough to detect the short-range forces then it is close enough to stick to the moisture. One way around this is tapping, or dynamic contact mode.

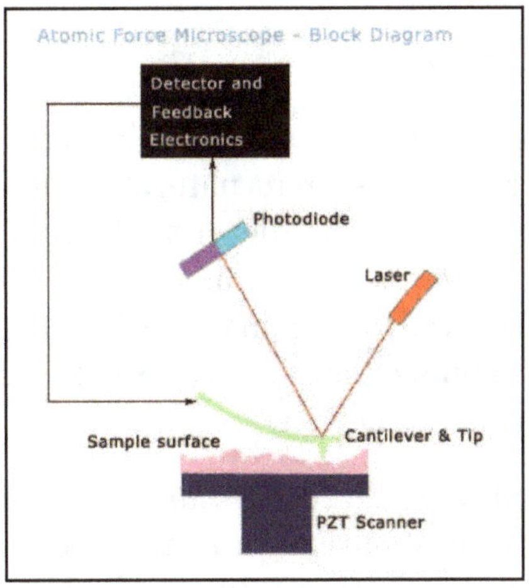

Tapping Mode

In tapping mode, the cantilever uses a piezoelectric element mounted on the top to oscillate it at near to its resonance frequency with an amplitude of up to 200nm.

The forces cause the amplitude to decrease as the tip gets close to the surface, and the height of the cantilever adjusts to keep the amplitude constant.

This tapping results in less damage to the sample than contact mode and is more accurate than non-contact mode when moisture is present on a sample.

Topographic Image

Image formation is a plotting method that produces a color mapping through changing the x-y position of the tip while scanning and recording the measured variable, i.e. the intensity of control signal, to each x-y coordinate. The color mapping shows the measured value corresponding to each coordinate. The image expresses the intensity of a value as a hue. Usually, the correspondence between the intensity of a value and a hue is shown as a color scale in the explanatory notes accompanying the image.

What is the Topographic Image of Atomic Force Microscope

Operation mode of image forming of the AFM are generally classified into two groups from the viewpoint whether it uses z-Feedback loop (not shown) to maintain the tip-sample distance to keep signal intensity exported by the detector. The first one (using z-Feedback loop), said to be "constant **XX** mode" (**XX** is something which kept by z-Feedback loop).

Topographic image formation mode is based on abovementioned "constant **XX** mode", z-Feedback loop controls the relative distance between the probe and the sample through outputting control signals to keep constant one of frequency, vibration and phase which typically corresponds to the motion of cantilever (for instance, voltage is applied to the Z-piezoelectric element and it moves the sample up and down towards the Z direction.

Details will be explained in the case that especially "constant df mode"(FM-AFM) among AFM as an instance in next section.

Topographic Image of FM-AFM

When the distance between the probe and the sample is brought to the range where atomic force may be detected, while a cantilever is excited in its natural eigen frequency (f_0), a phenomenon occurs that the resonance frequency (f) of the cantilever shifts from its original resonance frequency (natural eigen frequency). In other words, in the range where atomic force may be detected, the frequency shift ($df=f-f_0$) will be observed. So, when the distance between the probe and the sample is in the non-contact region, the frequency shift increases in negative direction as the distance between the probe and the sample gets smaller.

When the sample has concavity and convexity, the distance between the tip-apex and the sample varies in accordance with the concavity and convexity accompanied with a scan of the sample along x-y direction (without height regulation in z-direction). As a result, the frequency shift arises. The image in which the values of the frequency obtained by a raster scan along the x-y direction of the sample surface are plotted against the x-y coordination of each measurement point is called a constant-height image.

On the other hand, the df may be kept constant by moving the probe upward and downward in z-direction using a negative feedback (by using z-feedback loop) while the raster scan of the sample surface along the x-y direction. The image in which the amounts of the negative feedback (the moving distance of the probe upward and downward in z-direction) are plotted against the x-y coordination of each measurement point is a topographic image. In other words, the topographic image is a trace of the tip of the probe regulated so that the df is constant and it may also be considered to be a plot of a constant-height surface of the df.

Therefore, the topographic image of the AFM is not the exact surface morphology itself, but actually the image influenced by the bond-order between the probe and the sample, however, the topographic image of the AFM is considered to reflect the geographical shape of the surface more than the topographic image of a scanning tunnel microscope.

Force Spectroscopy

Another major application of AFM (besides imaging) is force spectroscopy, the direct measurement of tip-sample interaction forces as a function of the gap between the tip and sample (the result of this measurement is called a force-distance curve). For this method, the AFM tip is extended towards and retracted from the surface as the deflection of the cantilever is monitored as a function of piezoelectric displacement. These measurements have been used to measure nanoscale contacts, atomic bonding, Van der Waals forces, and Casimir forces, dissolution forces in liquids and single molecule stretching and rupture forces. Furthermore, AFM was used to measure, in an aqueous environment, the dispersion force due to polymer adsorbed on the substrate. Forces of the order of a few piconewtons can now be routinely measured with a vertical distance resolution of better than 0.1 nanometers. Force spectroscopy can be performed with either static or dynamic modes. In dynamic modes, information about the cantilever vibration is monitored in addition to the static deflection.

Problems with the technique include no direct measurement of the tip-sample separation and the common need for low-stiffness cantilevers, which tend to 'snap' to the surface. These problems are not insurmountable. An AFM that directly measures the tip-sample separation has been developed. The snap-in can be reduced by measuring in liquids or by using stiffer cantilevers, but in the latter case a more sensitive deflection sensor is needed. By applying a small dither to the tip, the stiffness (force gradient) of the bond can be measured as well.

Biological Applications and other

Force spectroscopy is used in biophysics to measure the mechanical properties. of living material (such as tissue or cells) or detect structures of different stiffness buried into the bulk of the sample using the stiffness tomography. Another application was to measure the interaction forces between from one hand a material stuck on the tip of the cantilever, and from another hand the surface of particles either free or occupied by the same material. From the adhesion force distribution curve, a mean value of the forces has been derived. It allowed to make a cartography of the surface of the particles, covered or not by the material.

Identification of Individual Surface Atoms

The AFM can be used to image and manipulate atoms and structures on a variety of surfaces. The atom at the apex of the tip "senses" individual atoms on the underlying surface when it forms incipient chemical bonds with each atom. Because these chemical interactions subtly alter the tip's vibration frequency, they can be detected and mapped. This principle was used to distinguish between atoms of silicon, tin and lead on an alloy surface, by comparing these 'atomic fingerprints' to values obtained from large-scale density functional theory (DFT) simulations.

The trick is to first measure these forces precisely for each type of atom expected in the sample, and then to compare with forces given by DFT simulations. The team found that the tip interacted most strongly with silicon atoms, and interacted 24% and 41% less strongly with tin and lead atoms, respectively. Thus, each different type of atom can be identified in the matrix as the tip is moved across the surface.

Probe

An AFM probe has a sharp tip on the free-swinging end of a cantilever that is protruding from a holder. The dimensions of the cantilever are in the scale of micrometers. The radius of the tip is usually on the scale of a few nanometers to a few tens of nanometers. (Specialized probes exist with much larger end radii, for example probes for indentation of soft materials.) The cantilever holder, also called holder chip – often 1.6 mm by 3.4 mm in size – allows the operator to hold the AFM cantilever/probe assembly with tweezers and fit it into the corresponding holder clips on the scanning head of the atomic force microscope.

This device is most commonly called an "AFM probe", but other names include "AFM tip" and "cantilever" (employing the name of a single part as the name of the whole device). An AFM probe is a particular type of SPM (scanning probe microscopy) probe.

AFM probes are manufactured with MEMS technology. Most AFM probes used are made from silicon (Si), but borosilicate glass and silicon nitride are also in use. AFM probes are considered consumables as they are often replaced when the tip apex becomes dull or contaminated or when the cantilever is broken. They can cost from a couple of tens of dollars up to hundreds of dollars per cantilever for the most specialized cantilever/probe combinations.

Just the tip is brought very close to the surface of the object under investigation, the cantilever is deflected by the interaction between the tip and the surface, which is what the AFM is designed to measure. A spatial map of the interaction can be made by measuring the deflection at many points on a 2D surface.

Several types of interaction can be detected. Depending on the interaction under investigation, the surface of the tip of the AFM probe needs to be modified with a coating. Among the coatings used are gold – for covalent bonding of biological molecules and the detection of their interaction with a surface, diamond for increased wear resistance and magnetic coatings for detecting the magnetic properties of the investigated surface. Another solution exists to achieve high resolution magnetic imaging : having the probe equip with a microSQUID. The AFM tips is fabricated using silicon micro machining and the precise positioning of the microSQUID loop is done by electron beam lithography.

The surface of the cantilevers can also be modified. These coatings are mostly applied in order to increase the reflectance of the cantilever and to improve the deflection signal.

Forces vs Tip Geometry

The forces between the tip and the sample strongly depend on the geometry of the tip. Various studies were exploited in the past years to write the forces as a function of the tip parameters.

Among the different forces between the tip and the sample, the water meniscus forces are highly interesting, both in air and liquid environment. Other forces must be considered, like the Coulomb force, van der Waals forces, double layer interactions, solvation forces, hydration and hydrophobic forces.

Water Meniscus

Water meniscus forces are highly interesting for AFM measurements in air. Due to the ambient humidity, a thin layer of water is formed between the tip and the sample during air measumements. The resulting capillary force gives rise to a strong attractive force that pulls the tip onto the surface. In fact, the adhesion force measured between tip and sample in ambient air of finite humidity is usually dominated by capillary forces. As a consequence, it is difficult to pull the tip away from the surface. For soft samples including many polymers and in particular biological materials, the strong adhesive capillary force gives rise to sample degradation and distruction upon imaging in contact mode. Historically, these problems were an important motivation for the development of dynamic imaging in air (e.g. 'tapping mode'). During tapping mode imaging in air, capillary bridges still form. Yet, for suitable imaging conditions, the capillary bridges are formed and broken in every oscillation cycle of the cantilever normal to the surface, as can be inferred from a analysis of cantilever amplitude and phase vs. distance curves. As a consequence, distructive shear forces are largely reduced and soft samples can be investigated.

In order to quantify the equilibrium capillary force, it is necessary to start from the Laplace equation for pressure:

$$P = \gamma_L \left(\frac{1}{r_1} + \frac{1}{r_0}\right) \simeq \frac{\gamma_L}{r_{eff}}$$

where γ_L is the surface energy and r_0 and r_1 are defined in the figure.

The pressure is applied on an area of,

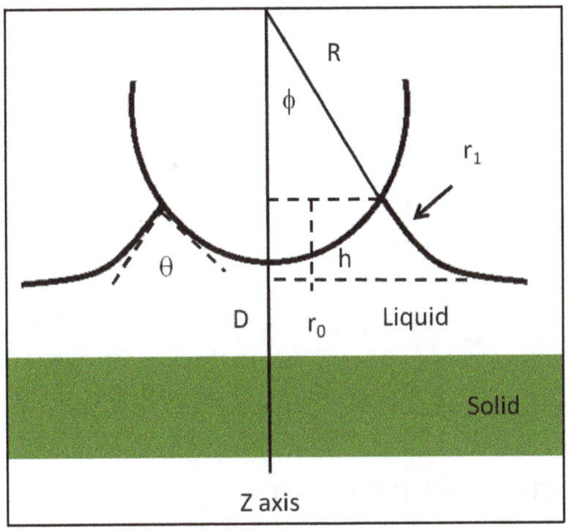

Model for AFM water meniscus.

$$A \simeq 2\pi R \simeq \left[reff\,(1+\cos\theta)+h\right]$$

where d, θ, and h are defined in the figure.

The force which pulles together the two surfaces is,

$$F = 2\pi R\gamma_L \left(1+\cos\theta+\frac{h}{r_{eff}}\right)$$

The same formula could also be calculated as a function of relative humidity.

Gao calculated formulas for different tip geometries. As an example, the forse decreases by 20% for a conical tip with respect to a spherical tip.

When these forces are calculated, a difference must be made between the wet on dry situation and the wet on wet situation.

For a spherical tip, the force is:

$$f_m = -2\pi R\gamma_L(\cos\theta + \cos\phi)\left(1 - \frac{dh}{dD}\right)$$

for dry on wet

$$f_m = -2\pi R\gamma_L \frac{dr_0}{dD}$$

for wet on wet

where θ is the contact angle of the dry sphere and φ is the immersed angle, as shown in the figure Also R,h and D are illustrated in the same figure.

For a conical tip, the formula becomes:

$$f_m = -2\pi R\gamma_L \frac{\tan\delta}{\cos\delta}(\cos\theta + \sin\delta)(hD)\left(1 - \frac{dh}{dD}\right)$$

for dry on wet

$$f_m = -2\pi R\gamma_L\left(\frac{1}{\cos\delta} + \sin\delta\right)(r_0)\left(\frac{dr_0}{dD}\right)$$

for wet on wet

where δ is the half cone angle and r_0 and h are parameters of the meniscus profile.

AFM Cantilever-deflection Measurement

Beam-deflection Measurement

The most common method for cantilever-deflection measurements is the beam-deflection method. In this method, laser light from a solid-state diode is reflected off the back of the cantilever and collected by a position-sensitive detector (PSD) consisting of two closely spaced photodiodes, whose output signal is collected by a differential amplifier. Angular displacement of the cantilever results in one photodiode collecting more light than the other photodiode, producing an output signal (the difference between the photodiode signals normalized by their sum), which is proportional to the deflection of the cantilever. The sensitivity of the beam-deflection method is very high, a noise floor on the order of 10 fm $Hz^{-1/2}$ can be obtained routinely in a well-designed system. Although this method is sometimes called the 'optical lever' method, the signal is not amplified if the beam path is made longer. A longer beam path increases the motion of the reflected spot on the photodiodes, but also widens the spot by the same amount due to diffraction, so that the same amount of optical power is moved from one photodiode to the other. The 'optical leverage' (output signal of the detector divided by deflection of the cantilever) is inversely proportional to the numerical aperture of the beam focusing optics, as long as the focused laser spot is small enough to fall completely on the cantilever. It is also inversely proportional to the length of the cantilever.

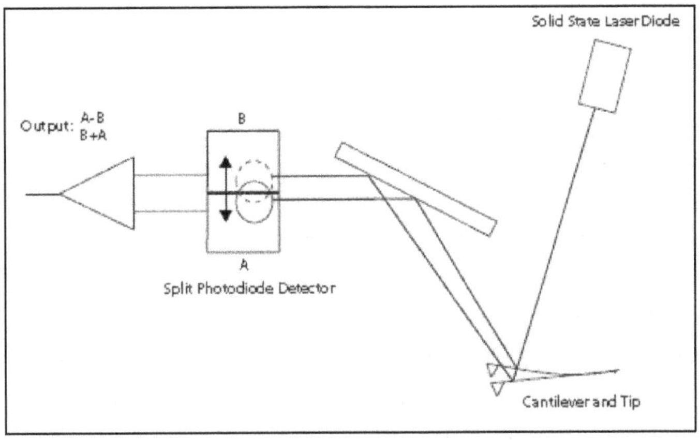

AFM beam-deflection detection.

The relative popularity of the beam-deflection method can be explained by its high sensitivity and simple operation, and by the fact that cantilevers do not require electrical contacts or other special treatments, and can therefore be fabricated relatively cheaply with sharp integrated tips.

Other Deflection-measurement Methods

Many other methods for beam-deflection measurements exist.

- Piezoelectric detection: Cantilevers made from quartz (such as the qPlus configuration), or other piezoelectric materials can directly detect deflection as an electrical signal. Cantilever oscillations down to 10pm have been detected with this method.
- Laser Doppler vibrometry: A laser Doppler vibrometer can be used to produce very accurate deflection measurements for an oscillating cantilever (thus is only used in non-contact mode). This method is expensive and is only used by relatively few groups.
- Scanning tunneling microscope (STM): The first atomic microscope used an STM complete with its own feedback mechanism to measure deflection. This method is very difficult to implement, and is slow to react to deflection changes compared to modern methods.
- Optical interferometry: Optical interferometry can be used to measure cantilever deflection. Due to the nanometre scale deflections measured in AFM, the interferometer is running in the sub-fringe regime, thus, any drift in laser power or wavelength has strong effects on the measurement. For these reasons optical interferometer measurements must be done with great care (for example using index matching fluids between optical fibre junctions), with very stable lasers. For these reasons optical interferometry is rarely used.

- Capacitive detection: Metal coated cantilevers can form a capacitor with another contact located behind the cantilever. Deflection changes the distance between the contacts and can be measured as a change in capacitance.
- Piezoresistive detection: Cantilevers can be fabricated with piezoresistive elements that act as a strain gauge. Using a Wheatstone bridge, strain in the AFM cantilever due to deflection can be measured. This is not commonly used in vacuum applications, as the piezoresistive detection dissipates energy from the system affecting Q of the resonance.

Piezoelectric Scanners

AFM scanners are made from piezoelectric material, which expands and contracts proportionally to an applied voltage. Whether they elongate or contract depends upon the polarity of the voltage applied. Traditionally the tip or sample is mounted on a 'tripod' of three piezo crystals, with each responsible for scanning in the x,y and z directions. In 1986, the same year as the AFM was invented, a new piezoelectric scanner, the tube scanner, was developed for use in STM. Later tube scanners were incorporated into AFMs. The tube scanner can move the sample in the x, y, and z directions using a single tube piezo with a single interior contact and four external contacts. An advantage of the tube scanner compared to the original tripod design, is better vibrational isolation, resulting from the higher resonant frequency of the single element construction, in combination with a low resonant frequency isolation stage. A disadvantage is that the x-y motion can cause unwanted z motion resulting in distortion. Another popular design for AFM scanners is the flexure stage, which uses separate piezos for each axis, and couples them through a flexure mechanism.

Scanners are characterized by their sensitivity, which is the ratio of piezo movement to piezo voltage, i.e., by how much the piezo material extends or contracts per applied volt. Because of differences in material or size, the sensitivity varies from scanner to scanner. Sensitivity varies non-linearly with respect to scan size. Piezo scanners exhibit more sensitivity at the end than at the beginning of a scan. This causes the forward and reverse scans to behave differently and display hysteresis between the two scan directions. This can be corrected by applying a non-linear voltage to the piezo electrodes to cause linear scanner movement and calibrating the scanner accordingly. One disadvantage of this approach is that it requires re-calibration because the precise non-linear voltage needed to correct non-linear movement will change as the piezo ages. This problem can be circumvented by adding a linear sensor to the sample stage or piezo stage to detect the true movement of the piezo. Deviations from ideal movement can be detected by the sensor and corrections applied to the piezo drive signal to correct for non-linear piezo movement. This design is known as a 'closed loop' AFM. Non-sensored piezo AFMs are referred to as 'open loop' AFMs.

The sensitivity of piezoelectric materials decreases exponentially with time. This causes most of the change in sensitivity to occur in the initial stages of the scanner's life. Piezoelectric scanners are run for approximately 48 hours before they are shipped from the factory so that they are past the point where they may have large changes in sensitivity. As the scanner ages, the sensitivity will change less with time and the scanner would seldom require recalibration, though various manufacturer manuals recommend monthly to semi-monthly calibration of open loop AFMs.

Advantages and Disadvantages

The first atomic force microscope.

Advantages

AFM has several advantages over the scanning electron microscope (SEM). Unlike the electron microscope, which provides a two-dimensional projection or a two-dimensional image of a sample, the AFM provides a three-dimensional surface profile. In addition, samples viewed by AFM do not require any special treatments (such as metal/carbon coatings) that would irreversibly change or damage the sample, and does not typically suffer from charging artifacts in the final image. While an electron microscope needs an expensive vacuum environment for proper operation, most AFM modes can work perfectly well in ambient air or even a liquid environment. This makes it possible to study biological macromolecules and even living organisms. In principle, AFM can provide higher resolution than SEM. It has been shown to give true atomic resolution in ultra-high vacuum (UHV) and, more recently, in liquid environments. High resolution AFM is comparable in resolution to scanning tunneling microscopy and transmission electron microscopy. AFM can also be combined with a variety of optical microscopy and spectroscopy techniques such as fluorescent microscopy of infrared spectroscopy, giving rise to scanning near-field optical microscopy, nano-FTIR and further expanding its applicability. Combined AFM-optical instruments have been applied primarily in the biological sciences but have recently attracted strong interest in photovoltaics and energy-storage research, polymer sciences, nanotechnology and even medical research.

Disadvantages

A disadvantage of AFM compared with the scanning electron microscope (SEM) is the single scan image size. In one pass, the SEM can image an area on the order of square millimeters with a depth of field on the order of millimeters, whereas the AFM can only image a maximum scanning area of about 150×150 micrometers and a maximum height on the order of 10-20 micrometers. One method of improving the scanned area size for AFM is by using parallel probes in a fashion similar to that of millipede data storage.

The scanning speed of an AFM is also a limitation. Traditionally, an AFM cannot scan images as fast as an SEM, requiring several minutes for a typical scan, while an SEM is capable of scanning at near real-time, although at relatively low quality. The relatively slow rate of scanning during AFM imaging often leads to thermal drift in the image making the AFM less suited for measuring accurate distances between topographical features on the image. However, several fast-acting designs were suggested to increase microscope scanning productivity including what is being termed videoAFM (reasonable quality images are being obtained with videoAFM at video rate: faster than the average SEM). To eliminate image distortions induced by thermal drift, several methods have been introduced.

AFM images can also be affected by nonlinearity, hysteresis, and creep of the piezoelectric material and cross-talk between the *x*, *y*, *z* axes that may require software enhancement and filtering. Such filtering could "flatten" out real topographical features. However, newer AFMs utilize real-time correction software (for example, feature-oriented scanning) or closed-loop scanners, which practically eliminate these problems. Some AFMs also use separated orthogonal scanners (as opposed to a single tube), which also serve to eliminate part of the cross-talk problems.

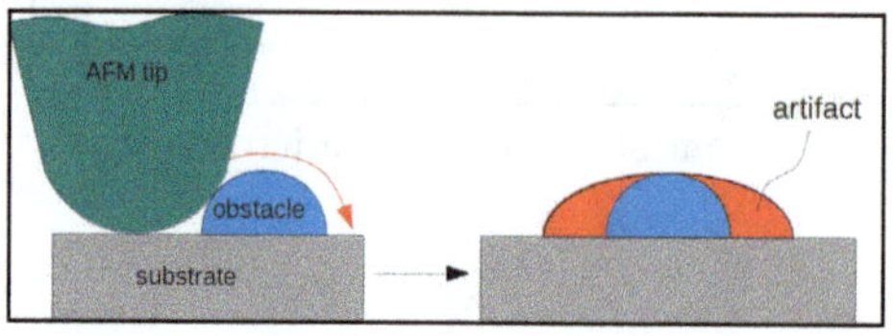

Showing an AFM artifact arising from a tip with a high radius of curvature with respect to the feature that is to be visualized.

As with any other imaging technique, there is the possibility of image artifacts, which could be induced by an unsuitable tip, a poor operating environment, or even by the sample itself, as depicted on the right. These image artifacts are unavoidable; however, their occurrence and effect on results can be reduced through various methods. Artifacts resulting from a too-coarse tip can be caused for example by inappropriate handling or de facto collisions with the sample by either scanning too fast or having an unreasonably rough surface, causing actual wearing of the tip.

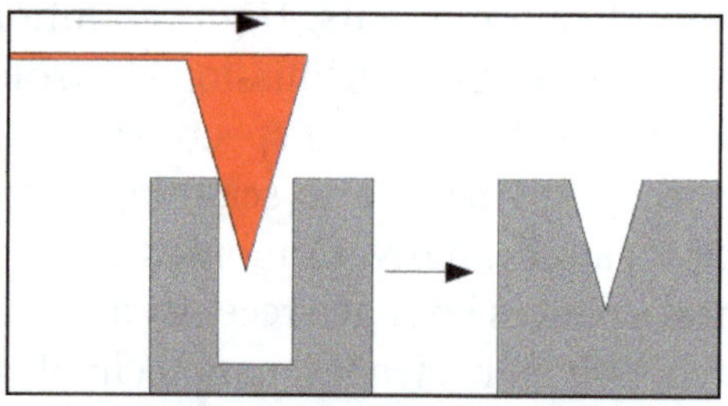

AFM artifact, steep sample topography.

Due to the nature of AFM probes, they cannot normally measure steep walls or overhangs. Specially made cantilevers and AFMs can be used to modulate the probe sideways as well as up and down (as with dynamic contact and non-contact modes) to measure sidewalls, at the cost of more expensive cantilevers, lower lateral resolution and additional artifacts.

Other Applications in Various Fields of Study

The latest efforts in integrating nanotechnology and biological research have been successful and show much promise for the future. Since nanoparticles are a potential vehicle of drug delivery, the biological responses of cells to these nanoparticles are continuously being explored to optimize their efficacy and how their design could be improved. Pyrgiotakis et al. were able to study the interaction between CeO_2 and Fe_2O_3 engineered nanoparticles and cells by attaching the engineered nanoparticles to the AFM tip. Studies have taken advantage of AFM to obtain further information on the behavior of live cells in biological media. Real-time atomic force spectroscopy (or nanoscopy) and dynamic atomic force spectroscopy have been used to study live cells and membrane proteins and their dynamic behavior at high resolution, on the nanoscale. Imaging and obtaining information on the topography and the properties of the cells has also given insight into chemical processes and mechanisms that occur through cell-cell interaction and interactions with other signaling molecules (ex. ligands). Evans and Calderwood used single cell force microscopy to study cell adhesion forces, bond kinetics/dynamic bond strength and its role in chemical processes such as cell signaling. Scheuring, Lévy, and Rigaud reviewed studies in which AFM to explore the crystal structure of membrane proteins of photosynthetic bacteria.Alsteen et al. have used AFM-based nanoscopy to perform a real-time analysis of the interaction between live mycobacteria and antimycobacterial drugs (specifically isoniazid, ethionamide, ethambutol, and streptomycine), which serves as an example of the more in-depth analysis of pathogen-drug interactions that can be done through AFM.

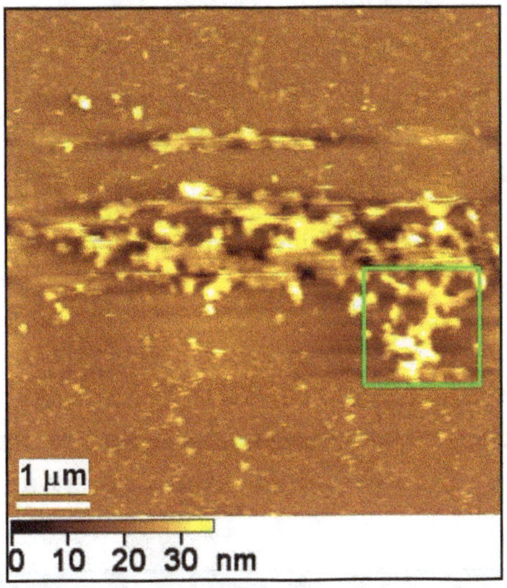

AFM image of part of a Golgi apparatus isolated from HeLa cells.

Scanning Tunneling Microscopy (STM)

Scanning tunneling microscope (STM) is a type of microscope whose principle of operation is based on the quantum mechanical phenomenon known as tunneling, in which the wavelike properties of electrons permit them to "tunnel" beyond the surface of a solid into regions of space that are forbidden to them under the rules of classical physics. The probability of finding such tunneling electrons decreases exponentially as the distance from the surface increases. The STM makes use of this extreme sensitivity to distance. The sharp tip of a tungsten needle is positioned a few angstroms from the sample surface. A small voltage is applied between the probe tip and the surface, causing electrons to tunnel across the gap. As the probe is scanned over the surface, it registers variations in the tunneling current, and this information can be processed to provide a topographical image of the surface.

The STM appeared in 1981, when Swiss physicists Gerd Binnig and Heinrich Rohrer set out to build a tool for studying the local conductivity of surfaces. Binnig and Rohrer chose the surface of gold for their first image. When the image was displayed on the screen of a television monitor, they saw rows of precisely spaced atoms and observed broad terraces separated by steps one atom in height. Binnig and Rohrer had discovered in the STM a simple method for creating a direct image of the atomic structure of surfaces. Their discovery opened a new era for surface science, and their impressive achievement was recognized with the award of the Nobel Prize for Physics in 1986.

The Quantum Corral

The STM image below shows the direction of standing-wave patterns in the local density of states of the Cu (111) surface. These spatial oscillations are quantum-mechanical interference patterns caused by scattering of the two-dimensional electron gas off the Fe atoms and point defects.

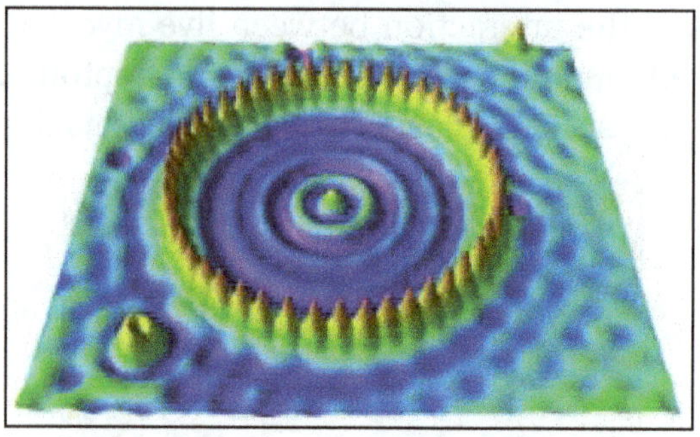

Courtesy of International Business Machines Corporation. Unauthorized use not permitted.

How an STM Works

The scanning tunneling microscope (STM) works by scanning a very sharp metal wire tip over a surface. By bringing the tip very close to the surface, and by applying an electrical voltage to the tip or sample, we can image the surface at an extremely small scale – down to resolving individual atoms.

3D rendered Scanning Tunneling Microscope image of atoms.

The STM is based on several principles. One is the quantum mechanical effect of tunneling. It is this effect that allows us to "see" the surface. Another principle is the piezoelectric effect. It is this effect that allows us to precisely scan the tip with angstrom-level control. Lastly, a feedback loop is required, which monitors the tunneling current and coordinates the current and the positioning of the tip. This is shown schematically below where the tunneling is from tip to surface with the tip rastering with piezoelectric positioning, with the feedback loop maintaining a current setpoint to generate a 3D image of the electronic topography.

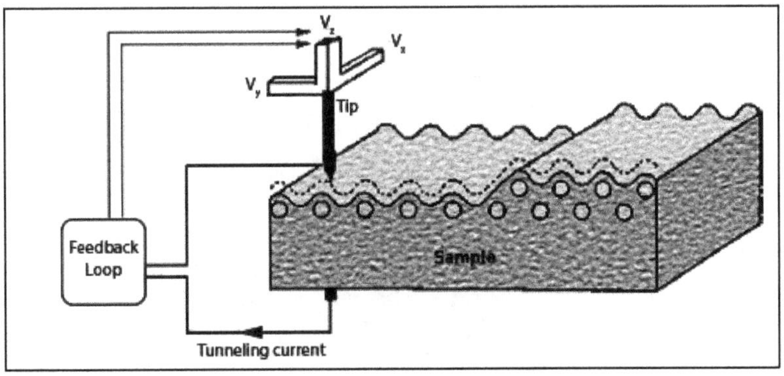

Schematic of scanning tunneling microscopy (STM).

Tunneling

Tunneling is a functioning concept that arises from quantum mechanics. Classically, an object hitting an impenetrable barrier will not pass through. In contrast, objects with a very small mass, such as the electron, have wavelike characteristics which permit such an event, referred to as tunneling.

Electrons behave as beams of energy, and in the presence of a potential $U(z)$, assuming 1-dimensional case, the energy levels $\psi_n(z)$ of the electrons are given by solutions to Schrödinger's equation,

$$-\frac{\hbar^2}{2m}\frac{\partial^2\psi_n(z)}{\partial z^2}+U(z)\psi_n(z)=E\psi_n(z)$$

where $\hbar$ is the reduced Planck's constant, z is the position, and m is the mass of an electron. If an electron of energy E is incident upon an energy barrier of height $U(z)$, the electron wave function is a traveling wave solution,

$$\psi_n(z)=\psi_n(0)e^{\pm ikz}$$

Where,

$$k=\frac{\sqrt{2m(E-U(z))}}{\hbar}$$

if E > U(z), which is true for a wave function inside the tip or inside the sample. Inside a barrier, E < U(z) so the wave functions which satisfy this are decaying waves,

$$\psi_n(z)=\psi_n(0)e^{\pm\kappa z}$$

Where,

$$\kappa=\frac{\sqrt{2m(U-E)}}{\hbar}$$

uantifies the decay of the wave inside the barrier, with the barrier in the +z direction for $-\kappa$.

Condensed matter experiments

ARPES
ACAR
Neutron scattering
X-ray spectroscopy
Quantum oscillations
Scanning tunneling microscopy

Knowing the wave function allows one to calculate the probability density for that electron to be found at some location. In the case of tunneling, the tip and sample wave functions overlap such that when under a bias, there is some finite probability to find the electron in the barrier region and even on the other side of the barrier. Let us assume the bias is V and the barrier width is W. This probability, P, that an electron at z=0 (left edge of barrier) can be found at $z=W$ (right edge of barrier) is proportional to the wave function squared,

$$P \propto |\psi_n(0)|^2 e^{-2\kappa W}.$$

If the bias is small, we can let $U - E \approx \varphi M$ in the expression for κ, where φM, the work function, gives the minimum energy needed to bring an electron from an occupied level, the highest of which is at the Fermi level (for metals at T=0 kelvins), to vacuum level. When a small bias V is applied to the system, only electronic states very near the Fermi level, within eV (a product of electron charge and voltage, not to be confused here with electronvolt unit), are excited. These excited electrons can tunnel across the barrier. In other words, tunneling occurs mainly with electrons of energies near the Fermi level.

A large scanning tunneling microscope, in the labs of the London Centre for Nanotechnology.

However, tunneling does require that there be an empty level of the same energy as the electron for the electron to tunnel into on the other side of the barrier. It is because of this restriction that the tunneling current can be related to the density of available or filled states in the sample. The current due to an applied voltage V (assume tunneling occurs sample to tip) depends on two factors: 1) the number of electrons between E_f and eV in the sample, and 2) the number among them which have corresponding free states to tunnel into on the other side of the barrier at the tip. The higher the density of available states the greater the tunneling current. When V is positive, electrons in the tip tunnel into empty states in the sample; for a negative bias, electrons tunnel out of occupied states in the sample into the tip.

Mathematically, this tunneling current is given by,

$$I \propto \sum_{E_f - eV}^{E_f} |\psi_n(0)|^2 \, e^{-2\kappa W}.$$

One can sum the probability over energies between $E_f - eV$ and E_f to get the number of states available in this energy range per unit volume, thereby finding the local density of states (LDOS) near the Fermi level. The LDOS near some energy E in an interval $\boldsymbol{\varepsilon}$ is given by,

$$\rho_s(z,E) = \frac{1}{\epsilon} \sum_{E-\epsilon}^{E} |\psi_n(z)|^2,$$

and the tunnel current at a small bias V is proportional to the LDOS near the Fermi level, which gives important information about the sample. It is desirable to use LDOS to express the current because this value does not change as the volume changes, while probability density does. Thus the tunneling current is given by,

$$I \propto V \rho_s(0, E_f) e^{-2\kappa W}$$

where $\rho_s(0,E_f)$ is the LDOS near the Fermi level of the sample at the sample surface. This current can also be expressed in terms of the LDOS near the Fermi level of the sample at the tip surface,

$$I \propto V \quad _s(W, E_f)$$

The exponential term in the above equations means that small variations in W greatly influence the tunnel current. If the separation is decreased by 1 Å, the current increases by an order of magnitude, and vice versa.

This approach fails to account for the *rate* at which electrons can pass the barrier. This rate should affect the tunnel current, so it can be treated using the Fermi's golden rule with the appropriate tunneling matrix element. John Bardeen solved this problem in his study of the metal-insulator-metal junction. He found that if he solved Schrödinger's equation for each side of the junction separately to obtain the wave functions ψ and χ for each electrode, he could obtain the tunnel matrix, M, from the overlap of these two wave functions.This can be applied to STM by making the electrodes the tip and sample, assigning ψ and χ as sample and tip wave functions, respectively, and evaluating M at some surface S between the metal electrodes, where z=0 at the sample surface and z=W at the tip surface.

Now, Fermi's Golden Rule gives the rate for electron transfer across the barrier, and is written

$$w = \frac{2\pi}{\hbar} | M |^2 \delta(E_\psi - E_\chi),$$

where $\delta(E_\psi - E_\chi)$ restricts tunneling to occur only between electron levels with the same energy. The tunnel matrix element, given by,

$$M = \frac{\hbar^2}{2m} \int_{z=z_0} (\chi^* \frac{\partial \psi}{\partial z} - \psi \frac{\partial \chi^*}{\partial z}) dS,$$

is a description of the lower energy associated with the interaction of wave functions at the overlap, also called the resonance energy.

Summing over all the states gives the tunneling current as,

$$I = \frac{4\pi e}{\hbar} \int_{-\infty}^{+\infty} \left[f\left(E_f - eV + e\right) - f\left(E_f + e\right) \right] \rho_S\left(E_f - eV + e\right) \rho_T\left(E_f + e\right) M |^2 \, de$$

where f is the Fermi function, ρ_S and ρ_T are the density of states in the sample and tip, respectively. The Fermi distribution function describes the filling of electron levels at a given temperature T.

Piezoelectric Effect

The piezoelectric effect was discovered by Pierre Curie in 1880. The effect is created by squeezing the sides of certain crystals, such as quartz or barium titanate. The result is the creation of opposite charges on the sides. The effect can be reversed as well; by applying a voltage across a piezoelectric crystal, it will elongate or compress.

These materials are used to scan the tip in an scanning tunneling microscopy (STM) and most other scanning probe techniques. A typical piezoelectric material used in scanning probe microscopy is PZT (lead zirconium titanate).

Feedback Loop

Electronics are needed to measure the current, scan the tip, and translate this information into a form that we can use for STM imaging. A feedback loop constantly monitors the tunneling current and makes adjustments to the tip to maintain a constant tunneling current. These adjustments are recorded by the computer and presented as an image in the STM software. Such a setup is called a constant current image.

In addition, for very flat surfaces, the feedback loop can be turned off and only the current is displayed. This is a constant height image.

Applications

Several surfaces have been studied with the STM. The arrangement of individual atoms on the metal surfaces of gold, platinum, nickel, and copper have all been accurately documented. The absorption and diffusion of different species such as oxygen and the epitaxial growth of gold on gold, silver on gold, and nickel on gold also have been examined in detail.

The surfaces of silicon have been studied more extensively than those of any other material. The surfaces are prepared by being heated in vacuum to temperatures so high that the atoms there rearrange their positions in a process called surface reconstruction. The reconstruction of the silicon surface designated (111) has been studied in minute detail. Such a surface reconstructs into an intricate and complex pattern known as the Takayanagi 7 × 7 structure. The position, the chemical reactivity, and the electronic configuration of each atomic site on the 7 × 7 surface has been measured with the STM. The reconstruction of the silicon surface designated (100) is more simple. The surface atoms form pairs, or dimers, that fit into rows that extend across the entire silicon surface.

"Vacuum tunneling" of electrons from tip to sample can take place even though the environment in the region surrounding the tip is not a vacuum but is filled with molecules of gas or liquids. With a tip-sample spacing as small as five angstroms, there is little room for molecules—even though they may exist in the surrounding atmosphere. The STM can operate in ambient atmosphere as well as in high vacuum. Indeed, it has been operated in air, in water, in insulating fluids, and in the ionic solutions used in electrochemistry. It is much more convenient than ultrahigh-vacuum instruments. When a high-vacuum environment is employed, its purpose is not to improve the performance of the STM but rather to ensure the cleanliness of the sample surface.

The STM can be cooled to temperatures less than 4 K (−269 °C, or −452 °F)—the temperature of liquid helium. It can be heated above 973 K (700 °C, or 1,300 °F). The low temperature is used to investigate the properties of superconducting materials, while the high temperature is employed to study the rapid diffusion of atoms across the surface of metals and their corrosion.

The STM is used primarily for imaging, but there are many other modalities that have been explored. The strong electric field between tip and sample has been utilized to move atoms along the sample surface. It has been used to enhance the etching rates in various gases. In one instance, a voltage of four volts was applied; the field at the tip was strong enough to remove atoms from the tip and deposit them on a substrate. This procedure has been employed with a gold tip to fabricate small gold islands or clusters on the substrate with several hundred atoms of gold in each cluster. These nanostructures are used to pattern the surface on a scale that is unprecedented.

Scanning Electron Microscopy (SEM)

A scanning electron microscope (SEM) scans a focused electron beam over a surface to create an image. The electrons in the beam interact with the sample, producing various signals that can be used to obtain information about the surface topography and composition

Why use Electrons Instead of Light in a Microscope

Given sufficient light, the human eye can distinguish two points 0.2 mm apart, without the aid of any additional lenses. This distance is called the resolving power or resolution of the eye. A lens or an assembly of lenses (a microscope) can be used to magnify this distance and enable the eye to see points even closer together than 0.2 mm.

A modern light microscope has a maximum magnification of about 1000x. The resolving power of the microscope was not only limited by the number and quality of the lenses but also by the wavelength of the light used for illumination. White light has wavelengths from 400 to 700 nanometers (nm). The average wavelength is 550 nm which results in a theoretical limit of resolution (not visibility) of the light microscope in white light of about 200 – 250 nm. The figure below shows two points at the limits of detection and the two individual spots can still be distinguished. The right image shows the two points so close together that the central spots overlap.

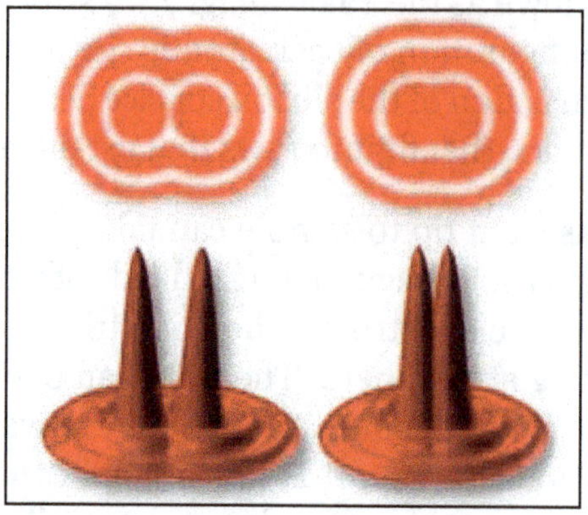

The electron microscope was developed when the wavelength became the limiting factor in light microscopes. Electrons have much shorter wavelengths, enabling better resolution.

Compare an Optical Microscope vs a Scanning Electron Microscope

As dimensions are shrinking for materials and devices, many structures can no longer be characterized by light microscopy. For example, to determine the integrity of a nanofiber layer for filtration, as shown here, electron microscopy is required to characterize the sample.

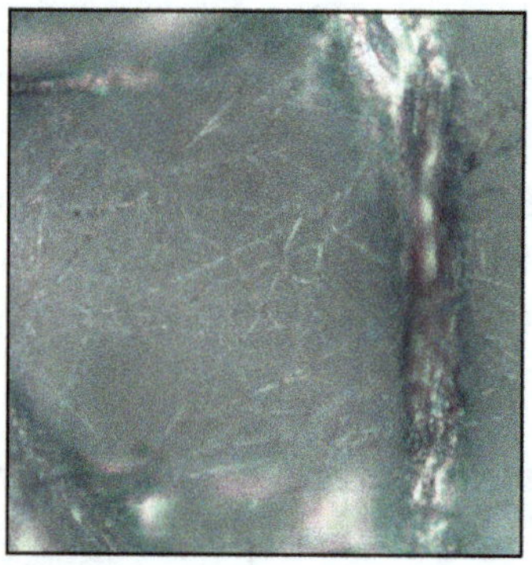
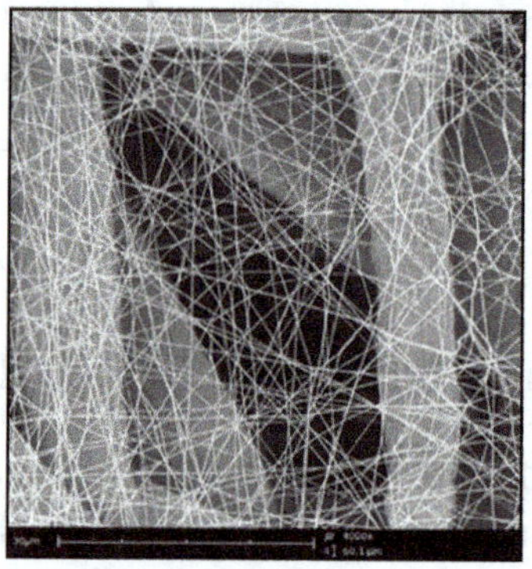

How a Scanning Electron Microscope Works

The main SEM components include:

- Source of electrons.
- Column down which electrons travel with electromagnetic lenses.
- Electron detector.
- Sample chamber.
- Computer and display to view the images.

Electrons are produced at the top of the column, accelerated down and passed through a combination of lenses and apertures to produce a focused beam of electrons which hits the surface of the sample. The sample is mounted on a stage in the chamber area and, unless the microscope is designed to operate at low vacuums, both the column and the chamber are evacuated by a combination of pumps. The level of the vacuum will depend on the design of the microscope.

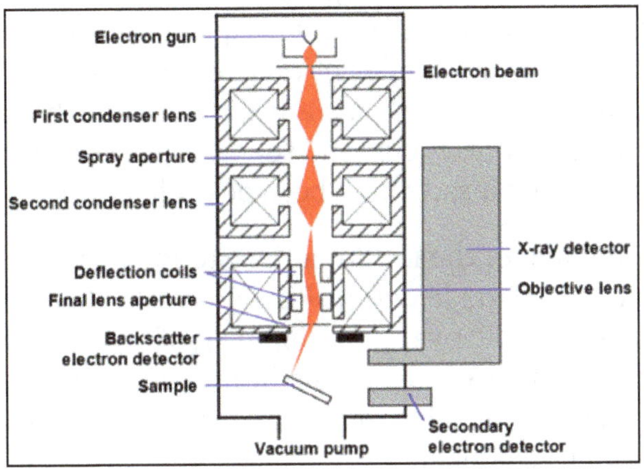

The position of the electron beam on the sample is controlled by scan coils situated above the objective lens. These coils allow the beam to be scanned over the surface of the sample. This beam rastering or scanning, as the name of the microscope suggests, enables information about a defined area on the sample to be collected. As a result of the electron-sample interaction, a number of signals are produced. These signals are then detected by appropriate detectors.

Sample-electron Interaction

The scanning electron microscope (SEM) produces images by scanning the sample with a high-energy beam of electrons. As the electrons interact with the sample, they produce secondary electrons, backscattered electrons, and characteristic X-rays. These signals are collected by one or more detectors to form images which are then displayed on the computer screen. When the electron beam hits the surface of the sample, it penetrates the sample to a depth of a few microns, depending on the accelerating voltage and the density of the sample. Many signals, like secondary electrons and X-rays, are produced as a result of this interaction inside the sample.

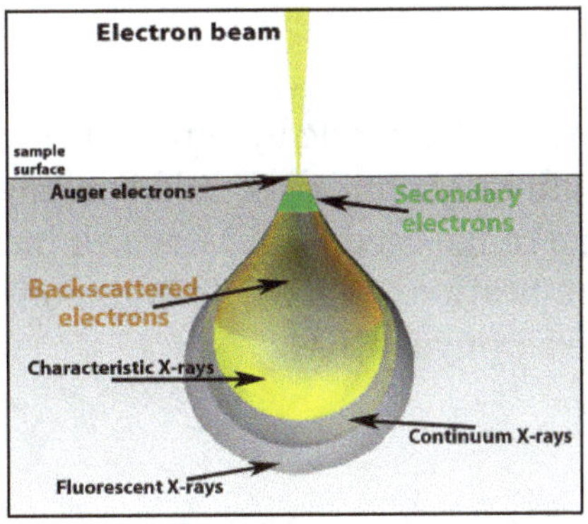

The maximum resolution obtained in an SEM depends on multiple factors, like the electron spot size and interaction volume of the electron beam with the sample. While it cannot provide atomic resolution, some SEMs can achieve resolution below 1 nm. Typically, modern full-sized SEMs provide resolution between 1-20 nm whereas desktop systems can provide a resolution of 20 nm or more.

Environmental SEM

Conventional SEM requires samples to be imaged under vacuum, because a gas atmosphere rapidly spreads and attenuates electron beams. As a consequence, samples that produce a significant amount of vapour, e.g. wet biological samples or oil-bearing rock, must be either dried or cryogenically frozen. Processes involving phase transitions, such as the drying of adhesives or melting of alloys, liquid transport, chemical reactions, and solid-air-gas systems, in general cannot be observed. Some observations of living insects have been possible however.

The first commercial development of the ESEM in the late 1980s allowed samples to be observed in low-pressure gaseous environments (e.g. 1–50 Torr or 0.1–6.7 kPa) and high relative humidity (up to 100%). This was made possible by the development of a secondary-electron detector capable of operating in the presence of water vapour and by the use of pressure-limiting apertures with differential pumping in the path of the electron beam to separate the vacuum region (around the gun and lenses) from the sample chamber.

The first commercial ESEMs were produced by the ElectroScan Corporation in USA in 1988. ElectroScan was taken over by Philips (who later sold their electron-optics division to FEI Company) in 1996.

ESEM is especially useful for non-metallic and biological materials because coating with carbon or gold is unnecessary. Uncoated Plastics and Elastomers can be routinely examined, as can uncoated biological samples. Coating can be difficult to reverse, may conceal small features on the surface of the sample and may reduce the value of the results obtained. X-ray analysis is difficult with a coating of a heavy metal, so carbon coatings are routinely used in conventional SEMs, but ESEM makes it possible to perform X-ray microanalysis on uncoated non-conductive specimens; however some specific for ESEM artifacts are introduced in X-ray analysis. ESEM may be the preferred for electron microscopy of unique samples from criminal or civil actions, where forensic analysis may need to be repeated by several different experts.

It is possible to study specimens in liquid with ESEM or with other liquid-phase electron microscopy methods.

Transmission SEM

The SEM can also be used in transmission mode by simply incorporating an appropriate detector below a thin specimen section. Both bright and dark field imaging has been reported in the generally low accelerating beam voltage range used in SEM, which increases the contrast of unstained biological specimens at high magnifications with a field emission electron gun. This mode of operation has been abbreviated by the acronym tSEM.

Color in SEM

Electron microscopes do not naturally produce color images, as an SEM produces a single value per pixel; this value corresponds to the number of electrons received by the detector during a small period of time of the scanning when the beam is targeted to the (x, y) pixel position.

This single number is usually represented, for each pixel, by a grey level, forming a "black-and-white" image. However, several ways have been used to get color electron microscopy images.

False Color using a Single Detector

- On compositional images of flat surfaces (typically BSE):

The easiest way to get color is to associate to this single number an arbitrary color, using a color look-up table (i.e. each grey level is replaced by a chosen color). This method is known as false color. On a BSE image, false color may be performed to better distinguish the various phases of the sample.

- On textured-surface images:

As an alternative to simply replacing each grey level by a color, a sample observed by an oblique beam allows researchers to create an approximative topography image. Such topography can then be processed by 3D-rendering algorithms for a more natural rendering of the surface texture.

SEM Image Coloring

Very often, published SEM images are artificially colored. This may be done for aesthetic effect, to clarify structure or to add a realistic appearance to the sample and generally does not add information about the specimen.

Coloring may be performed manually with photo-editing software, or semi-automatically with dedicated software using feature-detection or object-oriented segmentation.

Color Built using Multiple Electron Detectors

In some configurations more information is gathered per pixel, often by the use of multiple detectors.

As a common example, secondary electron and backscattered electron detectors are superimposed and a color is assigned to each of the images captured by each detector, with an end result of a combined color image where colors are related to the density of the components. This method is known as density-dependent color SEM (DDC-SEM). Micrographs produced by DDC-SEM retain topographical information, which is better captured by the secondary electrons detector and combine it to the information about density, obtained by the backscattered electron detector.

Analytical Signals Based on Generated Photons

Measurement of the energy of photons emitted from the specimen is a common method to get analytical capabilities. Examples are the energy-dispersive X-ray spectroscopy (EDS) detectors used in elemental analysis and cathodoluminescence microscope (CL) systems that analyse the intensity and spectrum of electron-induced luminescence in (for example) geological specimens. In SEM systems using these detectors it is common to color code these extra signals and superimpose them in a single color image, so that differences in the distribution of the various components of the specimen can be seen clearly and compared. Optionally, the standard secondary electron image can be merged with the one or more compositional channels, so that the specimen's structure and composition can be compared. Such images can be made while maintaining the full integrity of the original signal data, which is not modified in any way.

3D in SEM

SEMs do not naturally provide 3D images contrary to SPMs. However 3D data can be obtained using an SEM with different methods as follows.

3D SEM Reconstruction from a Stereo Pair

- photogrammetry is the most metrologically accurate method to bring the third dimension to SEM images. Contrary to photometric methods (next paragraph), photogrammetry calculates absolute heights using triangulation methods. The drawbacks are that it works only if there is a minimum texture, and it requires two images to be acquired from two different angles, which implies the use of a tilt stage. (Photogrammetry is a software operation that calculates the shift (or "disparity") for each pixel, between the left image and the right image of the same pair. Such disparity reflects the local height).

Photometric 3D SEM Reconstruction from a Four-quadrant Detector "Shape from Shading"

This method typically uses a four-quadrant BSE detector (alternatively for one manufacturer, a 3-segment detector). The microscope produces four images of the same specimen at the same time, so no tilt of the sample is required. The method gives metrological 3D dimensions as far as the slope of the specimen remains reasonable. Most SEM manufacturers now offer such built-in or optional four-quadrant BSE detector, together with proprietary software allowing to calculate a 3D image in real time.

Other approaches use more sophisticated (and sometimes GPU-intensive) methods like the optimal estimation algorithm and offer much better results at the cost of high demands on computing power.

In all instances, this approach works by integration of the slope, so vertical slopes and overhangs are ignored; for instance, if an entire sphere lies on a flat, little more than the upper hemisphere is seen emerging above the flat, resulting in wrong altitude of the sphere apex. The prominence of this effect depends on the angle of the BSE detectors with respect to the sample, but these detectors are usually situated around (and close to) the electron beam, so this effect is very common.

Photometric 3D Rendering from a Single SEM Image

This method requires an SEM image obtained in oblique low angle lighting. The grey-level is then interpreted as the slope, and the slope integrated to restore the specimen topography. This method is interesting for visual enhancement and the detection of the shape and position of objects ; however the vertical heights cannot usually be calibrated, contrary to other methods such as photogrammetry.

Other types of 3D SEm Reconstruction

- Inverse reconstruction using electron-material interactive models.
- Vertical stacks of SEM micrographs plus image-processing software.

- Multi-Resolution reconstruction using single 2D File: High-quality 3D imaging may be an ultimate solution for revealing the complexities of any porous media, but acquiring them is costly and time consuming. High-quality 2D SEM images, on the other hand, are widely available. Recently, a novel three-step, multiscale, multiresolution reconstruction method is presented that directly uses 2D images in order to develop 3D models. This method, based on a Shannon Entropy and conditional simulation, can be used for most of the available stationary materials and can build various stochastic 3D models just using a few thin sections.

Applications of 3D SEM

One possible application is measuring the roughness of ice crystals. This method can combine variable-pressure environmental SEM and the 3D capabilities of the SEM to measure roughness on individual ice crystal facets, convert it into a computer model and run further statistical analysis on the model. Other measurements include fractal dimension, examining fracture surface of metals, characterization of materials, corrosion measurement, and dimensional measurements at the nano scale (step height, volume, angle, flatness, bearing ratio, coplanarity, etc.).

Coordinate Measuring Machine (CMM)

Co-ordinate Measuring Machines are built rigidly and are very precise. They are equipped with digital readout or can be linked to computers for online inspection of parts. These machines can be placed close to machine tools for efficient inspection and rapid feedback for correction of processing parameter before the next part is made. They are also made more rugged to resist environmental effects in manufacturing plants such as temperature variations, vibration and dirt. Important features of the CMMs are:

- To give maximum rigidity to machines without excessive weight, all the moving members, the bridge structure, Z-axis carriage, and Z-column are made of hollow box construction.
- A map of systematic errors in machine is built up and fed into the computer system so that the error compensation is built up into the software.
- All machines are provided with their own computers with interactive dialogue facility and friendly software.
- Thermocouples are incorporated throughout the machine and interfaced with the computer to be used for compensation of temperature gradients and thus provide increased accuracy and repeatability.

A CMM consists of four main elements :

Main Structure

The machine incorporates the basic concept of three coordinate axes so that precise movement in x, y, and z directions is possible. Each axis is fitted with a linear measurement transducer. The transducers sense the direction of movement and gives digital display. Accordingly, there may be four types of arrangement:

Cantilever

The cantilever construction combines easy access and relatively small floor space requirements. It is typically limited to small and medium sized machines. Parts larger than the machine table can be inserted into the open side without inhibiting full machine travel. Figure shows a cantilever structure.

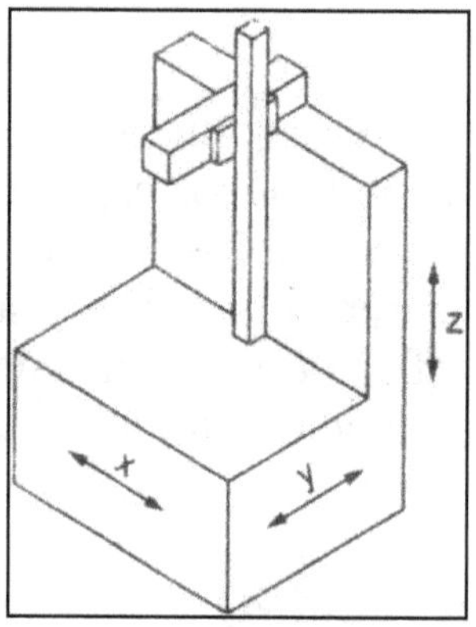

Cantilever Structure.

Bridge Type

The bridge arrangement over the table carries the quill (z-axis) along the x-axis and is sometimes referred to as a travelling bridge. It is claimed that the bridge construction provides better accuracy, although it may be offset by difficulty in making two members track in perfect alignment. This is by far the most popular CMM construction. Figure shows a bridge structure.

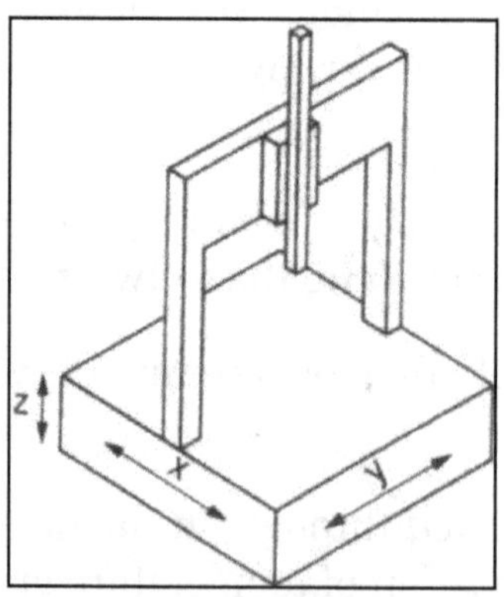

Bridge Structure.

Column Type

The column type machine is commonly referred to as a universal measuring machine rather than a CMM. These machines are usually considered gage room instruments rather than production floor machine. The direction of movements of the arms are as shown in figure. The constructional difference in column type with the cantilever type is with x and y-axes movements.

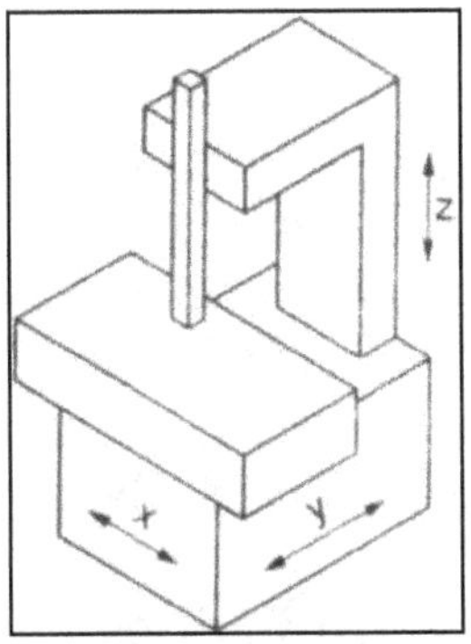

Column Structure.

Gantry

In a gantry type arrangement, arms are held by two fixed supports as shown in figure. Other two arms are capable of sliding over the supports.

Movements of the x, y and z-axes are also as shown in figure below. The gantry type construction is particularly suited for very large components and allows the operator to remain close to the area of inspection.

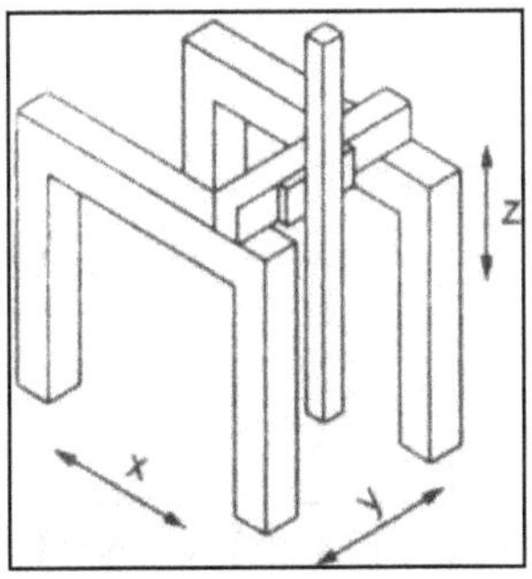

Gantry Structure.

Horizontal

Figure below shows the construction of a horizontal structure. The open structure of this arrangement provides optimum accessibility for large objects such as dies, models, and car bodies. Some horizontal arm machines are referred to as layout machines. There are some horizontal machines where the probe arm can rotate like a spindle to perform tramming operations. Tramming refers to accurate mechanical adjustment of instrument or machine with the help of tram.

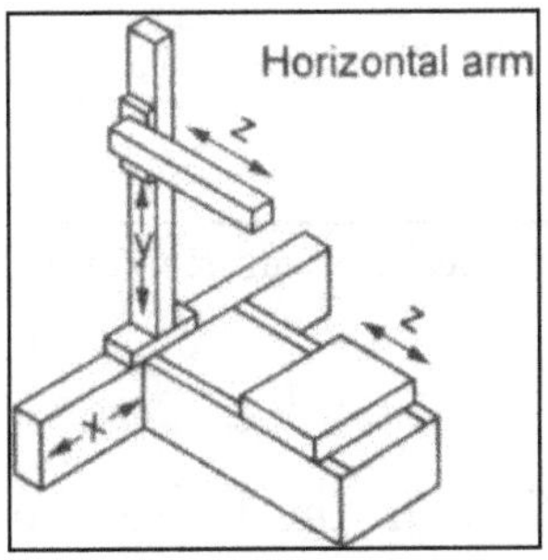

Horizontal Structure.

Probing System

It is the part of a CMM that sense the different parameters required for the calculation. Appropriate probes have to be selected and placed in the spindle of the CMM. Originally, the probes were solid or hard, such as tapered plugs for locating holes. These probes required manual manipulation to establish contact with the workpiece, at which time the digital display was read. Nowadays, transmission trigger-probes, optical transmission probes, multiple or cluster probes, and motorized probes are available. They are discussed in brief below:

Inductive and Optical Transmission Probes

Inductive and optical transmission probes have been developed for automatic tool changing. Power is transmitted using inductive linking between modules fitted to the machine structure and attached to the probe. figure below shows a schematic of the inductive transmission probe. The hard-wired transmission probe shown is primarily for tool setting and is mounted in a fixed position on the machine structure.

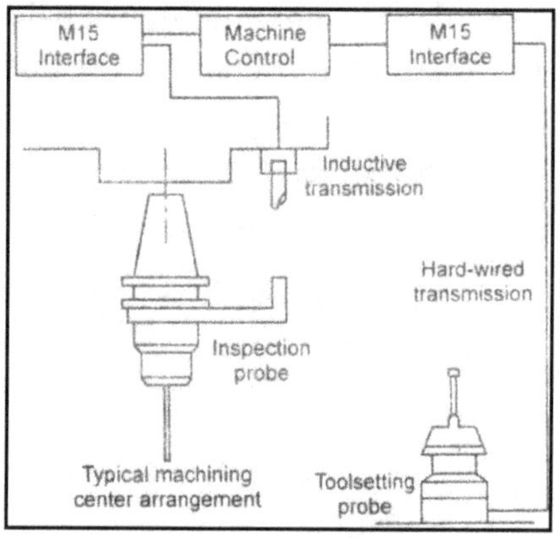

Inductive Probe System and Automatic Probe Changing.

The optical transmission probe shown in figure below allows probe rotation between gaging moves, making it particularly useful for datuming the probe. The wide-angle system allows greater axial movement of the probe and is suitable for the majority of installation.

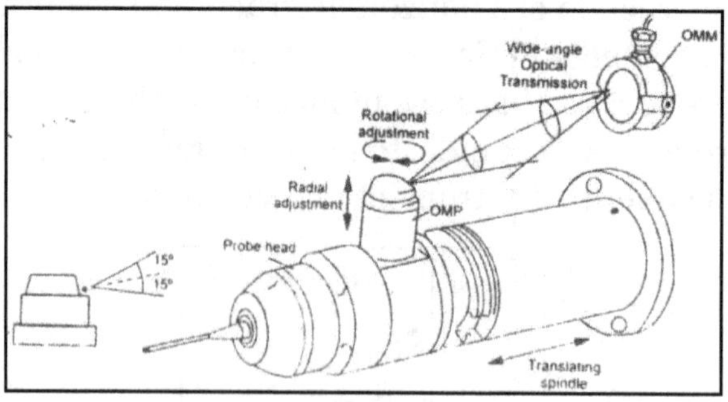

Optical Transmission Probe.

Motorized Probe

With the motorized probe, 48 positions in the horizontal axis, 15 in the vertical axis can be programmed for a total of 720 distinct probe orientations. Figure (b) below shows some typical applications for motorized probe. It shows that with a range of light weight extensions, the head can reach into deep holes and recesses. The second diagram shows that head of the probe is sufficiently compact to be regarded as an extension of the machine quill. This enables the inspection of complex components that would otherwise be impossible or involve complex setups.

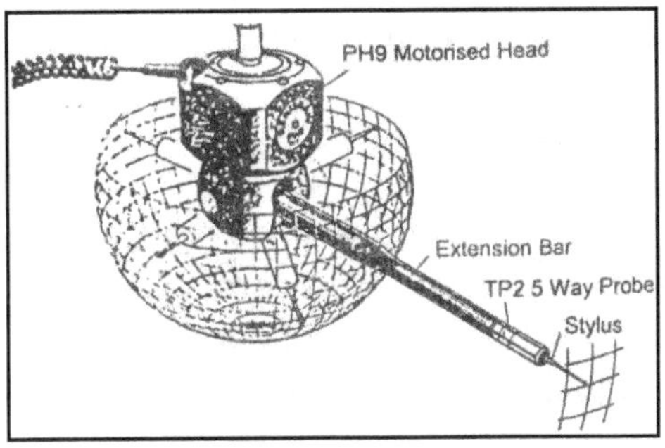

(a) Motorized Probe

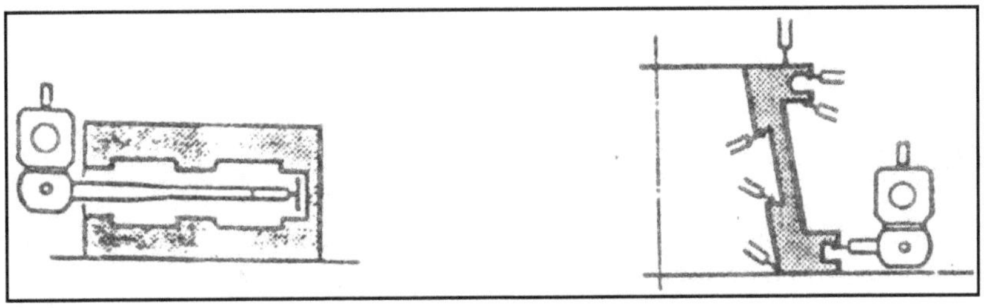

(b) Typical Applications of Motorized Probe

Multiple Styluses Probe Heads

Wide ranges of styli have been developed to suit many different gaging applications. Some of the different styli available are shown mounted on a multiple gaging head in figure below. The selection of stylus is done based on the application for which the probe is to be used.

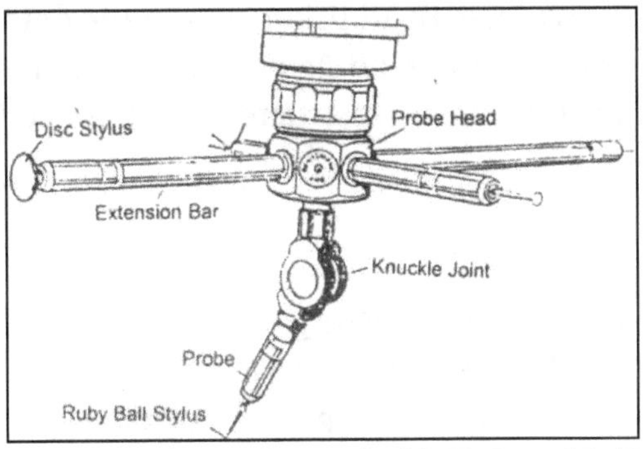

Multiple Stylus Probe Head with Variety of Styli.

Advantages of Cmm

CMM has got a number of advantages. The precision and accuracy given by a CMM is very high. It is because of the inherent characteristics of the measuring techniques used in CMM. Following are the main advantages that CMM can offer.

Flexibility

CMMs are essentially universal measuring machines and need not be dedicated to any particular task. They can measure almost any dimensional characteristic of a part configuration, including cams, gears and warped surfaces. No special fixtures or gages are required. Because probe contact is light, most parts can be inspected without being clamped to the table.

Reduced Setup Time

Part alignment and establishing appropriate reference points are very time consuming with conventional surface plate inspection techniques. Software allows the operator to define the orientation of the part on the CMM, and all subsequent data are corrected for misalignment between the parts-reference system and the machine coordinates.

Single Setup

Most parts can be inspected in a single setup, thus eliminating the need to reorient the parts for access to all features.

Improved Accuracy

All measurements in a CMM are taken from a common geometrically fixed measuring system, eliminating the introduction and the accumulation of errors that can result with hand-gage inspection methods and transfer techniques.

Reduced Operator Influence

The use of digital readouts eliminate the subjective interpretation of readings common with dial or vernier type measuring devices. Operator "feel" is virtually eliminated with modern touch-trigger probe systems, and most CMMs have routine measuring procedures for typical part features, such as bores or center distances. In computer assisted systems; the operator is under the control of a program that eliminates operator choice. In addition, automatic data recording, available on most machines, prevents errors in transcribing readings to the inspection report. This adds up to the fact that less skilled operators can be easily instructed to perform relatively complex inspection procedures.

References

- "Cornell researchers rotate a single molecule of oxygen, making a device that could be used for data storage | Cornell Chronicle". news.cornell.edu. Retrieved 09, August 2020
- Scanning-tunneling-microscope: britannica.com, Retrieved 17, July 2020
- G. Schitter; M. J. Rost (2008). "Scanning probe microscopy at video-rate". Materials Today. 11 (special issue): 40–48. doi:10.1016/S1369-7021(09)70006-9. ISSN 1369-7021. Archived from the original (PDF) on 2009-09-09
- Atomic-force-microscope: microscopemaster.com, Retrieved 28, May 2020
- K. Oura; V. G. Lifshits; A. A. Saranin; A. V. Zotov & M. Katayama (2003). Surface science: an introduction. Berlin: Springer-Verlag. ISBN 978-3-540-00545-2

- Scanning-electron-microscopy, techniques: nanoscience.com, Retrieved 30, June 2020
- Suzuki, E. (2002). "High-resolution scanning electron microscopy of immunogold-labelled cells by the use of thin plasma coating of osmium". Journal of Microscopy. 208 (3): 153–157. doi:10.1046/j.1365-2818.2002.01082.x
- Scanning-tunneling-microscope-285634: britannica.com Retrieved 20, April 2020
- R. V. Lapshin; O. V. Obyedkov (1993). "Fast-acting piezoactuator and digital feedback loop for scanning tunneling microscopes" (PDF). Review of Scientific Instruments. 64 (10): 2883–2887. Bibcode:1993RScI...64.2883L. doi:10.1063/1.1144377
- Corral, scanning-tunneling-microscopy: nanoscience.com, Retrieved 15, May 2020
- R. V. Lapshin (2011). "Feature-oriented scanning probe microscopy". In H. S. Nalwa. Encyclopedia of Nanoscience and Nanotechnology (PDF). 14. USA: American Scientific Publishers. pp. 105–115. ISBN 978-1-58883-163-7

Permissions

All chapters in this book are published with permission under the Creative Commons Attribution Share Alike License or equivalent. Every chapter published in this book has been scrutinized by our experts. Their significance has been extensively debated. The topics covered herein carry significant information for a comprehensive understanding. They may even be implemented as practical applications or may be referred to as a beginning point for further studies.

We would like to thank the editorial team for lending their expertise to make the book truly unique. They have played a crucial role in the development of this book. Without their invaluable contributions this book wouldn't have been possible. They have made vital efforts to compile up to date information on the varied aspects of this subject to make this book a valuable addition to the collection of many professionals and students.

This book was conceptualized with the vision of imparting up-to-date and integrated information in this field. To ensure the same, a matchless editorial board was set up. Every individual on the board went through rigorous rounds of assessment to prove their worth. After which they invested a large part of their time researching and compiling the most relevant data for our readers.

The editorial board has been involved in producing this book since its inception. They have spent rigorous hours researching and exploring the diverse topics which have resulted in the successful publishing of this book. They have passed on their knowledge of decades through this book. To expedite this challenging task, the publisher supported the team at every step. A small team of assistant editors was also appointed to further simplify the editing procedure and attain best results for the readers.

Apart from the editorial board, the designing team has also invested a significant amount of their time in understanding the subject and creating the most relevant covers. They scrutinized every image to scout for the most suitable representation of the subject and create an appropriate cover for the book.

The publishing team has been an ardent support to the editorial, designing and production team. Their endless efforts to recruit the best for this project, has resulted in the accomplishment of this book. They are a veteran in the field of academics and their pool of knowledge is as vast as their experience in printing. Their expertise and guidance has proved useful at every step. Their uncompromising quality standards have made this book an exceptional effort. Their encouragement from time to time has been an inspiration for everyone.

The publisher and the editorial board hope that this book will prove to be a valuable piece of knowledge for students, practitioners and scholars across the globe.

Index

www.ingramcontent.com/pod-product-compliance
Lightning Source LLC
LaVergne TN
LVHW080728210926
848485LV00003B/536
9781639893089